D1357912

Economics of Pollution

THE CHARLES C. MOSKOWITZ LECTURES NUMBER XI

KENNETH E. BOULDING

PROFESSOR OF ECONOMICS

UNIVERSITY OF COLORADO

ELVIS J. STAHR

PRESIDENT

NATIONAL AUDUBON SOCIETY

SOLOMON FABRICANT

PROFESSOR OF ECONOMICS

NEW YORK UNIVERSITY

FORMER DIRECTOR

NATIONAL BUREAU OF ECONOMIC RESEARCH

MARTIN R. GAINSBRUGH

ADJUNCT PROFESSOR OF ECONOMICS

NEW YORK UNIVERSITY

CHIEF ECONOMIST

NATIONAL INDUSTRIAL CONFERENCE BOARD

Economics of Pollution

THE CHARLES C. MOSKOWITZ LECTURES
SCHOOL OF COMMERCE
NEW YORK UNIVERSITY

NEW YORK *New York University Press* 1971

Library of Congress Catalog Card Number: 70-179-973
SBN: 8147-0967-2
Manufactured in the United States of America

FOREWORD

THE CHARLES C. MOSKOWITZ LECTURES, held at the College of Business and Public Administration of New York University, aim at advancing public understanding of the issues that are of major concern to business and the nation. Established through the generosity of Mr. Charles C. Moskowitz, a distinguished alumnus of the School of Commerce and formerly Vice President-Treasurer and a Director of Loew's, Inc., the lectures have enabled the School to make a significant contribution to public discussion and understanding of important issues affecting the American economy and its business enterprises.

The 1971 lectures series, held on March twenty-second, was on the overall theme *Economic Growth and the Problem of Environmental Pollution.* It brought to our platform two lecturers of outstanding reputation, one a world-famed economist and the other an able administrator of long experience who has for several years been devoted to maintaining environmental quality. The economist was Kenneth E. Boulding, Professor of Economics at the University of Colorado and a past president of the American Economic Association. The administrator was Elvis J. Stahr, now President of the National Audubon Society and formerly a university president and government official.

Professor Boulding delivered a paper on the topic "What Do Economic Indicators Indicate?: Quality and Quantity in the GNP." Mr. Stahr spoke about "Antipollution Policies, Their Nature and Their Impact on Corporate Profits." Sharing the platform with these eminent lecturers were two distinguished economists who are members of the faculty at New York University. Our University participants were Professor Solomon Fabricant, a noted scholar who has been long associated with The National Bureau of Economic Research, and Martin R. Gainsbrugh, Adjunct Professor of Economics and Chief Economist at The National Industrial Conference Board.

Professor Boulding's paper is an intellectual *tour de force,* which brings out with exceptional clarity the extraordinary difficulties of evaluation in deciding between "bigger" and "better." His attack on the problem begins with a discussion of the cognition of human beings of the world about them. Thus, he considers first the theory of quantity and quality, and only secondly the real meaning of national income statistics. Examining the evaluative complexities involved in judging quality and quantity, he points out that all of us perceive the world, that is, we form images of the world and evaluate those images, as a result of several processes. These processes begin when, as a newly born infant, we receive undifferentiated stimulii, or "white noise," from outside; we are endowed with a primitive value system of genetic origin that enables us to manifest some preferences. Immediately, however, the exogenous inputs begin to interact with the genetic value system and create in the infant human being an internal image of the outside world. This internal image becomes increasingly complex with age and experience, is imbued with preferences and values, and finally comprises "folk knowledge," that is, images of the immediate environment and evaluations of them.

But man is more than mere animal, and he has always sought to extend his image of the world and his evaluations of it. In this effort he has moved from magic and superstition to science. Scientific knowledge, however, involves a feedback, testing, and evaluation process of much greater subtlety and complexity than does that of folk knowledge. In particular, the former requires the use of statistical "indicators" as the "cues" through which we attempt to perceive realistic images of complex and total systems, such as the social or the economic system. However, statistical "indicators" range from simple numerical indicators to complex profiles, representing a hierarchy of rising "epistemological significance." Professor Boulding illustrated this hierarchy by moving from an absolute number, as a case of the simplest indicator, to evaluative indices, which are complex indicators whose movements are related to judgments of "better" or "worse." The intermediary levels of epistemological significance in the statistical indicators, from the simplest to the most complex, are the ratio, the proportion, rank order, syndromes, dynamic patterns in the space-time four-dimensional continuum (such as celestial mechanics), and "models" (that is, organized patterns of abstract relationships having dynamic patterns that are perceived to correspond to the dynamic syndromes in space-time). The unhappy fact is that, as Professor Boulding pointed out, much social conflict arises out of the different evaluation indices (functions) different individuals hold. However, he believes that it may be possible to break the overall system of society down into subsystems in which the evaluation function can be defined with a greater degree of certainty, and with respect to which we may have somewhat more evaluative certainty; that is, we might be able to decide whether things are going from bad to better, or from bad to worse. Examples are exhaustion of resources, capital deterioration (both physical and human), pollution, and the development of perverse dynamic processes (for example, arms races). Identifying these as "deteriorating systems," Professor Boulding pointed out that they result from failures of human learning and failures of the "payoff" system (that is, the rewarding of vice and the penalizing of virtue).

Before applying the foregoing principles to economic indicators (national income statistics), Boulding mentioned the importance of uncertainty in connection with the making of decisions based upon epistemological structures. He cautioned his listeners against confusing statistical and epistemological significance, although they are related through the concept of uncertainty. In particular, he warned of the real danger of elaborate "staff" analysis which, on the basis of simulation and gaming, creates an illusion of certainty for the powerful decision-maker in a situation which is really uncertain. As he says, "Under conditions of uncertainty, he who hesitates is saved . . . (but), zeroing in almost always leads to hitting the target of disaster."

Having set forth the foregoing principles regarding perception and a hierarchy of epistemological significance, Professor Boulding turned to the national income statistics, the lessons we possibly could learn from them, and their defects. Using the concept of "Gross Capacity Product" in current dollars (that is, the GNP with zero unemployment), he undertook an examination of major components of the GCP, and possible offsetting changes in them. Into that examination he intertwined evaluative concepts of welfare, or of "good" or "bad," but always with a profound appreciation of the judgmental and hence uncertain aspects of the analysis. Thus, he commented "If we were to deduct pollution, education, health, and commuting, as costs, from the Per Capita Disposible Income, the rise in the last forty years might look very modest indeed. It is very doubtful whether Americans are as much as twice as rich as their grandfathers, which is clearly a modest rate of real economic growth." Or, ". . . the relatively small magnitude of the Vietnam War, even by comparison with the Korean War . . . should warn us to be careful in the qualitative interpretation of these quantitative phenomena, for the impact of the Vietnam War on American society has been very much larger qualitatively than was that of the Korean War, in spite of the fact that quantitatively it is not as large a proportion of the economy."

Moving up the scale of complexity in his hierarchy of epistemological significance, Professor Boulding pointed out that "a genuinely cybernetic apparatus in the economy would see to it that a decline in one item would set forces into play to produce increase in other items." He added that we do not now have anything like this for small depressions, although we probably do have something like it for large ones. However, to acquire knowledge of the operation of the economic system, for example as regards fluctuations in the level of economic activity, it is necessary to have a model of the system. Neither Presidents Herbert Hoover nor

Franklin D. Roosevelt had one, and they were consequently unsuccessful in combatting the Great Depression. However, Keynes' equilibrium model has helped. But even it is inadequate, and Boulding speculated that, despite our present "Keynesian based" cybernetic machinery for dealing with economic downturns, it is conceivable that there could be a repetition of the Great Depression. On the other side, he noted that our success in the last three decades in preventing severe depressions has taught us nothing about how to achieve full employment without inflation.

All the considerations pointed out above brought Boulding and his audience finally to the most difficult problem of all: the evaluation of economic systems. And here the lecturer showed great skepticism of the certainty displayed by critics and moralists, who, as he puts it, "In the Name of loving mankind, . . . actually despise the tastes of the vulgar majority and wish to impose their own, of course more refined, tastes on mankind as a whole." Thus, instead of answering for all of us the question of "bigger versus better," he sought to expand our insights into the complexities of the matter, and so to move us a bit ahead in the search for the answer. And in this he succeeded.

Elvis J. Stahr's lecture was somewhat less esoteric than that given by Boulding, and it did not present quite the same exquisite intellectual face. In terms of Boulding's hierarchy of epistemological significance, it was made at the level of a subsystem of society. But, as Boulding pointed out, that is the level at which the evaluation function can be defined with a greater degree of certainty. Thus, Stahr's discussion began with a consideration of our deteriorating environment, moving on to: (1) the nature of the pollution problems; (2) the nature of antipollution policy; (3) the choice of instruments for antipollution policy; and (4) the burdens of antipollution programs, to which he appended a note on pesticides before stating his conclusions. However, while a deteriorating environment is in Boulding's *schema* or hierarchy of epistemological significance at the level of a subsystem, it is in Stahr's hierarchy placed higher. Indeed, he spoke of the ecosystem, that is, "the web of life" or "balance of nature." And it is clearly his opinion that fundamental and continuing deterioration of *that system* endangers all human systems, whether social, economic, political, or, indeed, physiological.

Identifying the explosion of population and the explosion of technology as the two primary sources of environmental deterioration, Stahr, perhaps unwittingly, highlighted the evaluation problems so basic to deciding issues of "better" or "worse." There is readily available today the technical knowledge to control population growth, but there is no unanimity of values as to its application. So some speak about family planning while others speak about enlarging total output and achieving greater "justice" in the distribution of that output. Of course, the latter are not so likely to take into account the ecological consequences of their actions. For example, *The New York Times* of May 22, 1971 told of a bitter struggle by a group of Navajo and Hopi Indians in the Southwest who are fighting a huge complex of coal-fired power plants. The plants were originally planned in the late fifties and early sixties, in response to the hugh population growth in the region, and the demand for electrical energy to power the appliances of modern civilization. But these plants are belching smoke into once-clear skies and polluting them. Should the region's natural attributes be preserved in unsullied beauty? If so, then what would that mean for the region's ability to maintain its present, let alone its projected, population? The answers are neither simple nor easy. In any event, Stahr's paper did not plead for a stop to economic development, but rather for the inclusion of ecologists in its planning.

Moving into the details of his argument, Stahr pointed out that the nature of the contemporary pollution problem centers on what he called ecosystem overload, which is a condition in which natural processes cannot recycle all of the waste substances dumped into the environment. The result is a buildup of wastes and, if unchecked, ultimate destruction of the environment *and its capacity to sustain life.*

Additionally, the environment is a free dumping ground. It is a "commons," and no individual user has an incentive to protect it.

How are we to develop an antipollution policy which will prevent ecosystem overload? Stahr proposed three principles as necessary underpinnings for such a policy: (1) to examine carefully the alternatives, that is, the choices, available to polluters, such as recycling of wastes, technological advances which substitute less polluting products and production processes for more polluting ones, and, if necessary, curbing the rate of growth in consumption (there's a neat evaluation issue as between pro-consumptionists and anti-polluters); (2) to discover the least costly ways of cleaning up the environment, because the costs of reversing the deterioration which has already occurred in the ecosystem will be large; and (3) the establishment of a "safe minimum standard" of environmental purity, rather than the pursuit of an ideal which would involve "excessive" sacrifice of consumable goods. The goal would be "to eliminate pollution when the physical and aesthetic discomfort it creates, and its damage to people and things, are more costly than the value of the goods whose production or use has caused the pollution." But, in Boulding's context, whose discomfort and whose estimate of values are involved?

On the basis of the foregoing principles, Stahr presented several instruments for antipollution policy: (1) government standards for products; (2)

government standards for production processes; (3) taxes on pollutant emissions or substances; and (4) subsidies for control of pollution through tax rebates. Reviewing these instruments, Stahr spoke strongly in favor of the third, although the other ones have been the most heavily relied upon to this point. Indeed that is why Stahr is disenchanted with them, for he finds the record unsatisfactory. And he believes the tax approach will provide incentives to polluters to act to reduce pollution. To the degree pollution occurs, he wants it to be expensive. It is his opinion that, by making it expensive, the polluters will have incentives not to pollute, or, if they continue to pollute, government will have funds to finance public corrective action.

Stahr does not think the costs of cleaning up the environment, to a safe minimum standard, will be small. He accepts the range of estimates from 50 to 100 billion dollars during the next five years. However, he believes there are some offsets which should be included in any cost benefit comparison, that is, (1) the cost of present pollution; (2) the productive aspects of many antipollution remedies (for example, by-product recovery); and (3) who pays, in connection with which he presented a fairly extensive review of tax incidence and shifting as well as some comments favorable to the imposition of sumptuary taxes upon products that cause pollution. He added the judgment that, based on such evidence as is presently available, the likelihood of business failures and adverse competitive impacts in international trade is not great. Citing Kneese's estimates with approval, he noted that the extreme limit in the increase in production cost due to strict waste control programs was 10 percent, while the most likely amount for most industries would be 5 percent. Of course these costs will finally be reflected in the nation's GNP, which will be somewhat smaller, as well as in its unemployment, which will be somewhat greater, and its domestic income, which will be smaller. Stahr does not think international trade and our balance of payments position would be adversely affected to a serious degree. In all these estimates he relies on economic models used by Robert Anderson, Allen V. Kneese, Wassily Leontief, and Ralph d'Arge. And, in Boulding's terms, he attaches epistemological significance to the statistics produced by the models. As for those plants and firms which will die, "mortality has always been a fact of life."

Stahr concluded his paper with some words on the problem of pesticides, synthetic chemical substances used in agriculture and elsewhere to control plant pests. But his words were strong and condemnatory, although he did not argue for the complete elimination of pesticides. In short, he seems to think that at present they do more damage than good, and cutbacks in their use could be achieved without threatening the total output of some particular crop *everywhere*. As to local or regional damage to crop output, why "mortality has always been a fact of life." In his words, "This is very close to a situation where almost no one except the chemical industry and the adversely affected farm regions are worse off and most of us are better off, and it is to this that the conservation movement has been directing its efforts on this problem."

In concluding, Stahr did not make any sweeping claims. He thinks a cleaner environment, if achieved through a combination of intelligent policies which emphasize antipollution incentives but which do not ignore regulation and other approaches, will mean somewhat less output and a somewhat different mix of goods and services for us to consume. But, in his opinion, "The costs are worth undertaking," and he sees no reason for "paralyzing fear" over revolutionary consequences.

Noting that the very important issues raised by the discussion would not be immediately settled, Solomon Fabricant pointed out that "the problem of environmental pollution is . . . raised by life itself. . . . More specifically, human life means garbage and sewage and exhaled air. And the greater the number of human beings and the higher the standard of living attained by them, the greater tends to be the volume of undesired by-products of production and consumption. . . ." Like Stahr, Fabricant sees the population and technology explosions as roots of the problem. More, a projection of present growth rates in these two areas would mean a rate of environmental pollution so enormous as to be almost beyond comprehension. But the direction of causation is not only from economic growth to pollution. There is a feedback, and we may expect it to bring about adjustments to meet the problem of environmental pollution. However, there is no room for complacency, and Fabricant quotes Justice Holmes to the effect that "for the inevitable to happen takes some sweat." Specifically, it takes much knowledge about the facts of production and consumption, as well as about the costs of alternative actions and about "externalities," that is, the good or bad effects on people of transactions to which they are not a party.

With respect to policy, Fabricant indicated these points were important: (1) expenditures for antipollution will not reduce real income, so long as such expenditures yield benefits exceeding the already existent costs of pollution; (2) while the costs of antipollution in the long run are borne by consumers, in the short run they may fall on producers and considerations of equity may warrant subsidies in some cases; (3) since some consumers prefer the consumption of goods and services which produce more pollutants than others, those consumers should have to pay relatively higher prices for the consumption of those items; (4) the adoption of user charges for scarce environmental services; and (5) the exercise of caution, gradualness, and experimentation in adopting policies, with the understanding that caution is no excuse for inaction.

Finally, Fabricant observed the difficult political problem inherent in the fact that different people have different values and view national objectives differently. As for life itself, why living systems mutate as well as reproduce, and it is conceivable that life will continue on the earth—even if the earth becomes a cesspool. But would it be human life? In any case, mutation which permits life in a cesspool is not the answer. It is rather the adaptation of our habits and our institutions concerning population and the other factors relevant to the problem of environmental pollution. And on that score Fabricant is optimistic.

Martin R. Gainsbrugh's comments were more limited in scope than the remarks of the other participants, for he centered his attention more directly on the national accounts and GNP concepts. In this connection, he invited our attention to this question: Should the heavy investment in environmental control be treated as other capital investment and charged off over a long time period, or treated as a current social cost? The answer involves net national product and profits. And profits are related to new investment and the level of economic activity. These points are not meant to imply a negative attitude with respect to action against pollution, but rather a concern for the need to take appropriate tax actions so that antipollution efforts will not have adverse economic consequences which outweigh their benefits.

Beyond the foregoing, Gainsbrugh felt it to be important to say a few words about the worth of the GNP measures. They have enabled us to comprehend the workings of the economic system far better than before, and, consequently, to design and apply policies which have greatly enhanced economic growth and better living standards. As to the pollution attendant on growth, Gainsbrugh looks toward the development of a separate social account which will serve as a factual basis for the development of antipollution policies as useful as the economic policies which have been based on our economic accounts.

As Fabricant thought, we obtained no final answers. But the nature of the problem was defined with somewhat greater sharpness, and there appeared to be a fundamental consensus (with some differences of degree) as to its seriousness. There was also significant consensus on several points, most notably the importance of value differences between people and the probable superiority of antipollution policies involving incentives rather than regulations. If the lectures have moved us from the state of "folk knowledge" to that of science, as Boulding might put it, then they have been worthwhile and have served their purpose. And on that score I am optimistic.

Abraham L. Gitlow

THE CHARLES C. MOSKOWITZ LECTURES

THE CHARLES C. MOSKOWITZ LECTURES were established through the generosity of a distinguished alumnus of the School of Commerce, Mr. Charles C. Moskowitz of the Class of 1914, who retired after many years as Vice President-Treasurer and a Director of Loew's, Inc.

In establishing these lectures, it was Mr. Moskowitz's aim to contribute to the understanding of the function of business and its underlying disciplines in society by providing a public forum for the dissemination of enlightened business theories and practices.

The School of Commerce and New York University are deeply grateful to Mr. Moskowitz for his continued interest in, and contribution to, the educational and public service program of his alma mater.

This volume is the eleventh in the Moskowitz series. The earlier ones were:

February, 1961 *Business Survival in the Sixties*
Thomas F. Patton, President and Chief Executive Officer
Republic Steel Corporation

November, 1961 *The Challenges Facing Management*
Don G. Mitchell, President
General Telephone and Electronics Corporation

November, 1962 *Competitive Private Enterprise Under Government Regulation*
Malcolm A. MacIntyre, President
Eastern Air Lines

November, 1963 *The Common Market: Friend or Competitor?*
Jesse W. Markham, Professor of Economics, Princeton University
Charles E. Fiero, Vice President, The Chase Manhattan Bank
Howard S. Piquet, Senior Specialist in International Economics, Legislative Reference Service, The Library of Congress

November, 1964 *The Forces Influencing the American Economy*

Jules Backman, Research Professor of Economics, New York University

Martin R. Gainsbrugh, Chief Economist and Vice President, National Industrial Conference Board

November, 1965 *The American Market of the Future*

Arno H. Johnson, Vice President and Senior Economist, J. Walter Thompson Company

Gilbert E. Jones, President, IBM World Trade Corporation

Darrell B. Lucas, Professor of Marketing and Chairman of the Department, New York University

November, 1966 *Government Wage-Price Guideposts in the American Economy*

George Meany, President, American Federation of Labor and Congress of Industrial Organizations

Roger M. Blough, Chairman of the Board and Chief Executive Officer, United States Steel Corporation

Neil H. Jacoby, Dean, Graduate School of Business Administration, University of California at Los Angeles

November, 1967 *The Defense Sector in the American Economy*

Jacob K. Javits, United States Senator, New York

Charles J. Hitch, President, University of California

Arthur F. Burns, Chairman, Federal Reserve Board

November, 1968 *The Urban Environment: How It Can Be Improved*

William E. Zisch, Vice-chairman of the Board, Aerojet-General Corporation

Paul H. Douglas, Chairman, National Commission on Urban Problems

Professor of Economics, New School for Social Research

Robert C. Weaver, President, Bernard M. Baruch College of the City University of New York

Former Secretary of Housing and Urban Development

November, 1969 *Inflation: The Problems It Creates and the Policies It Requires*

Arthur M Okun, Senior Fellow, The Brookings Institution

Henry H. Fowler, General Partner, Goldman, Sachs & Co.

Milton Gilbert, Economic Adviser, Bank for International Settlements

WHAT DO ECONOMIC INDICATORS INDICATE?: QUALITY AND QUANTITY IN THE GNP

Kenneth E. Boulding

I. *The Theory of Quantity and Quality*

WHEN WE CONTRAST quantity with quality we are suggesting that the bigger is not necessarily the better. For a baby, growth in weight is evidently desirable; for the adult, it simply means that he is turning into fat. For the poor, growth in income is entirely desirable; for the rich, it may simply mean corruption and luxury. My own output of words has been considerably increased by the use of the dictaphone, but I worry a little bit about what has been happening to the sense that they make.

The innocent words "quantity" and "quality" are two bushy tails each attached to an enormous bear, one of which is labeled "bigger" and the other "better." The "bigger" problem is that of what do indicators really indicate. If an indicator goes "up," is there something in the real world, whatever that is, that has become larger? The "better" problem is the problem of evaluative systems, and values, human or otherwise, have been a thorn in the flesh for economists and philosophers for a long time.

Evaluation is characteristic of all living things, even down to the amoeba, which prefers food to a piece of grit. We might even extend the concept down to the elements themselves and say that carbon "prefers" holding hands with four hydrogen atoms rather than with three, but this is perhaps stretching the concept rather far. It is much more than mere analogy, however, to say that all living things have values, that is, they order certain parts of their environment on a scale of better or worse. As we go up in the scale of life, the field over which the ordering is made becomes more and more complex as the image, that is, the cognitive structure or the pattern inside the nervous system which corresponds to the patterns outside, also becomes more complex. Man, however, is the only organism in this part of the universe that has the effrontery to put a scale of better or worse over practically the whole universe. When evaluations of different people differ, as they generally do, new problems arise. If *A* thinks that state of the universe *X* is better than *Y*, and *B* thinks that *Y* is better than *X*, at the very least they may argue about it. They may also fight about it, and if some decision has to be made which involves choice between *X* and *Y*, there will have to be some political mechanism by which this decision is reached.

The problem of this paper, therefore, is a very large one, but there is really no way of making it smaller. We cannot discuss the relations of quality and quantity without considering the whole process by which we form images of the world and the process by which we evaluate these images and compare and resolve differences in these evaluations. There are at least three processes involved here, which are frequently successive, though they are constantly interrelated. We all begin as a new-born baby who begins immediately to receive large inputs of information through his senses, which at first appears as a great big buzzing confusion, practically equivalent to undifferentiated "white noise." We also, however, come into the world endowed with a primitive value system of genetic origin, built into the nervous system of the new-born baby by its genetic code. This may not go much beyond preference for certain temperatures—dislike for being too hot or too cold, preference for being dry rather than wet, full rather than hungry, milk rather than water, and being cuddled rather than abandoned. Immediately, however, our information inputs begin to interact with the genetic value system and perhaps with internally generated information to build up an internal image of the outside world by processes of filtering, information input, and feedback, related to some sort of output. The baby cries and gets attention. A double process then begins of building up increasingly complex internal images, and also of developing complex preferences and values. We come to believe that the messages coming into us from outside originate in a "real world," which is largely independent of our own images of it, insofar as these images are a part of it. We suppose that our internal image comes increasingly to correspond to that of a real world outside us because, if it does not, we get into trouble, whereas, if it does, we get rewarded. We learn to operate in a world of space and time because, when our behavior is guided by images which do not correspond to the world around us, we fall down and hurt ourselves. We learn the language of those around us because our efforts in this direction are rewarded, whereas random noises are not. At all points, as our image of the real world grows in complexity, our evaluations of it likewise grow. We begin to like some people and dislike others, and to like some foods and dislike others, and so on.

These processes develop what I have called "folk knowledge" and "folk values," that is, the images of our immediate environments and our evaluations of these. These images and values constantly interact throughout our whole life. We see these processes, indeed, going on in a lecture room. The lecturer is producing a stream of successive words in his nervous system and is projecting these in the form of sound waves towards his hearers. His eyes and his ears are receiving information all the time from the room around him, which he is constantly interpreting in the light of what has been going on, as he recalls it, in the last few minutes. If he makes what he thinks is a joke and nobody laughs, he revises his evaluation either of the speech or of the audience. The skilled lecturer sizes up his audience, plays on them constantly, almost like a musical instrument, adapts what he is saying continuously to the feedback which he gets, operating all the time under an evaluation of the continuing process in which he tries to go from bad to better rather than from bad to worse. The audience likewise is making evaluations all the time and may either be enjoying itself or wishing it had not come, and looking anxiously at its watches. Everybody is constantly picking up cues from his environment and modifying either his image of the environment itself or his evaluation of it in the light of the cues received.

Man has never been content with images and evaluations of his immediate environment. He has always been curious about the larger environment, about the stars, the weather, animals and plants, and his own subtle and incomprehensible processes, and he has always sought to extend his image of the world and his evaluations out towards the universe, moving through magic and superstition, speculative philosophy and mythology, into poetry, religion, and science. The epistemological processes by which we acquire scientific knowledge are not essentially different from those by which we acquire folk knowledge, but the images themselves are much more complex and the processes by which we evaluate them are likewise more complex. As we move from folk knowledge to science, we continue to learn by a process of feedback, testing, and evaluation, but the images of the world which we form are more subtle and complex. The inferences that we draw from these, for instance, by mathematical or logical inference, are more rigid and exact, and in our information input we have to move from the "cues" which are adequate for folk knowledge, to "indicators" which are derived from some sort of instrumentation, either physical or social, and which are usually expressed in the form of a number, or a succession of numbers. These numbers are usually derived by statistical compilations of large masses of information, which are then filtered and reduced to some single number as an index or indicator. It is one of the paradoxes of science that the cues from which much of our knowledge is derived, such as facial expressions, tones of voice, complex retinal images and so on, are much more subtle and multidimensional than indicators. In the case of systems, however, such as the economy or the polity, the electrons and atoms of the physical world, or the cells and genes of living organisms, with which we do not have immediate personal contact, the multidimensional and structured cues with which we are familiar in ordinary life are not only difficult to obtain, but frequently are misleading when they are obtained. Hence, we have to rely on one-dimensional indicators, or at least a series of these, in order to achieve any realistic image of complex and total systems, such as the social system. It is perhaps significant that we describe a complex of indicators as a "profile," that is, a two-dimensional representation of a face. Even in our most ambitious moments, we never think of meeting the economic system, for instance, face to face and perceiving it directly in its enormous multiformity, multidimensionality, and complexity. We see it only in a glass, darkly, through indicators.

All sciences use indicators. They are certainly not peculiar to the social sciences. A chemist probes the structure of a compound with a spectroscope or chromotograph. He never really meets it face to face. A doctor, facing the enormous complexity of the human body, takes its temperature, blood pressure, and other indicators, and out of this diagnoses disease. The economist takes the Gross National Product, the price level, and deduces from these something about the economy, which, again, he never meets face to face. Cues, indeed, produce what might be called a "folk image" of the economy, which can often be very misleading. In 1970, for instance, the United States economy suffered a recession which statistically, to judge by the various economic indicators, was no worse, let us say, than that of 1958 or 1961. It seems to have produced, however, an enormous amount of folk lore derived from personal cues, such as news about friends being laid off, departments abolished, factories shut down, and so on, which may be very poor samples of the total reality, but which nevertheless may set up chain reactions which intensify both the image and the reality of economic decline.

We can postulate a certain hierarchy of rising "epistemological significance" as we move from simple numerical indicators towards complex profiles. The simplest indicator is an absolute number which represents a total sum or aggregate of something. Thus, we estimate the world's population at about 3.5 billion people, and we estimate the Gross National Product of the United States to be roughly a trillion dollars. In some cases, like a population figure, or the numbers of automobiles or bathtubs, the individual unit is significant object. Hence, the absolute number is a count of distinctive items in a set, each element of which has some significance in itself. In the case of a number like the Gross National Product, the unit is essentially arbitrary, even though it is derived from a long historical process. If we counted the Gross National Product in cents, it would be a hundred times larger than if we counted it in dollars, but though the number would be different, the quantity would be the same. Similarly, a length can be measured in feet or in meters, and can be represented indeed by any number whatever if we choose the unit appropriately. In all these cases, therefore, the actual number which constitutes the indicator derives its significance from the nature of the unit in which it is counted.

Absolute numbers mean something to us simply because we have experience with sets of different sizes. We can visualize fairly easily what it means to have five things or seven things on a table. Very large numbers are much harder to visualize or interpret, and in order to get an image of their significance we usually have to translate them into areas or volumes. Thus, most people find it very hard to visualize a billion. If we think of it as the number of inch cubes in a cube of 1000 inches, or about 83 feet a side, which would be about the size of a large concert hall, we can get at least some perceptual image. Similarly, we could visualize the whole population of the world standing in an area about 10 or 11 miles square.

The second level of significance is the *ratio.* Ratios are usually more significant than absolute numbers. Indeed, attempts to visualize absolute numbers often involve reducing them to some kind of ratio. Ratios are used to express figures like population density per square mile, GNP per capita, crop production per acre, persons per automobile, and so on. Ratios are frequently significant because they enable us to visualize large aggregate quantities in terms with which we are familiar. Thus, it is very hard to visualize what a trillion dollars Gross National Product means. If, however, we express this as a little under 5000 dollars per head, the figure comes within normal comprehension, simply because most of us can express our own income, the spending of which we are highly familiar, as a simple proportion of the Gross National Product per capita. We know what it means roughly in terms of commodities to have 5000 dollars a year. All kinds of statistics, therefore, become much more epistemologically significant when they are expressed in per capita terms.

The third level, a variant of ratio, is *proportion*, which again frequently adds to the significance of the indicators. Proportion implies that there is significant aggregate or totality—this is the "100 percent"—which can be broken down into significant components, so that we have a set of percentages which add up to 100 percent. Thus, the proportions of a population by age, sex, race, religion, income group, educational achievement, and so on are of great significance in interpreting the structure of a population. Sometimes a structural proportional relationship can be expressed in terms of a single statistic, like the Gini Index for inequality in the distribution of income. More often, however, no single index could possibly express the complexities of the structure, and we end up with something that looks like a profile. In the second part of this paper, I will show some significant insights which can be obtained by breaking up various aggregates of national income statistics into their components, and expressing these as percentages of the total.

The fourth level of epistemological significance which may apply to all three of the previous levels is that of rank order. Even absolute figures, like the GNP or population, become much more significant when they are compared with those of other countries or other societies, and they are arranged in some kind of rank order. The rank order of absolute GNP has considerable significance in assessing the relative power structure in the international system, even though the most significant rank order consists of that produced by the GNP, modified by some kind of coefficient of the "will to exercise power." Per capita GNP is at least a very rough measure of how rich a country is, although, as we shall see, it is a very imperfect measure of welfare, and here again, the rank order is of considerable interest.

A fifth level of epistemological significance is the perception of syndromes, that is, things that go together. There are many statistical tests of syndrome properties, such as Chi square, rank-order correlation, coefficient of correlation, and so on. The perception of syndromes is also of great importance in the definition of categories. Pigeon holes should contain, if not pigeons, at least birds of a feather, that is, elements which are either alike or which tend to go together in nature. Categorization still remains something of an art, especially in the social sciences, where it is a much more difficult problem than it is in the physical sciences, or even in the biological sciences. Thus, there is not much doubt about the categorization of the chemical elements. There are only marginal areas of doubt about the definitions of living species, but there is a great amount of doubt and argument about the categories of the social sciences. We can recall the famous "empty boxes" controversy among economists in the 1920s,[1] when the economic historian J. H. Clapman accused the Marshallian economists of having devised a lot of categories, such as industries of increasing return, industries of decreasing return, and so on, which did not correspond to any kind of empirical syndromes. There is a similar controversy over the Parsonian categories is sociology, which are regarded by many as empty boxes.

It is indeed one of the great tasks of science to perceive both similarities and differences which are not immediately apparent. It takes a quite sophisticated chemistry, for instance, to reveal that diamonds and graphite are exactly the same element, differing only in the spatial arrangements of the atoms. Categorization in the social sciences is much

1. J. H. Clapham, "Of Empty Economic Boxes," *The Economic Journal* (1922), p. 305. The various articles of the controversy are reprinted in Stigler and Boulding (eds.), *Readings in Price Theory* (Homewood, Ill.: Richard D. Irwin, 1952), for the American Economic Association.

more difficult and complex. Attempts have been made to identify personality types, for instance, or to identify cultural syndromes, but we still have a long way to go in this aspect of the social sciences. Nevertheless, what might be called the "cross-sectional syndrome" (Jamaicans are black, Swedes are white) is a very important element in our image of the world of society.

The sixth level is the development of the perception of dynamic patterns, which represent, as it were, syndromes in the space-time, four-dimensional continuum. The simplest example of this is celestial mechanics, which enables us to predict the movement of the planets and other heavenly bodies. Because we have kept careful records of their position in the sky over a large number of successive dates, we have been able to perceive relatively simple time patterns in their movement, which can be expressed, for instance, in the Laplacean equations. Another example would be the record of the simple growth of compound interest in a bank. These are what I have called "difference systems" because their essential characteristic is a stable relationship between successive states, especially as expressed in stable differences between "today" and "yesterday." Thus, in the case of simple exponential growth, the quantity today is a constant proportion of the quantity yesterday. The relationship may extend over a number of successive states, this number defining the relationship. Thus, if the condition today is a stable function of the condition yesterday and the day before, we have a relationship of the second degree.

Still another dynamic syndrome is the "creode," the pattern of development of a living organism from the fertilized egg to maturity and eventual death. We perceive a great stability in these patterns, so that we always expect kittens to grow up into cats, and puppies into dogs. When we perceive these four-dimensional patterns we are getting very close to understanding the "real world," but we see here also the dilemma that, even in complex systems, it is much easier to perceive cues, like the kitten, than it is to discover reliable indicators. It took the human race about 5000 years to ferret out the patterns of the solar system, even then only because the states of the solar system can be described easily in rather simple numbers. The subtle patterns of the total social system are both much more difficult to reduce to any simple indicator and much more difficult to perceive and describe.

The seventh level of epistemological significance is the development of "models," that is, organized patterns of abstract relationships having dynamic properties that are perceived to correspond to the dynamic syndromes in space-time, or perhaps, at a simpler level, to correspond to the cross-sectional patterns which are descriptive of the state of the world at a moment of time. The atomic theory of molecules, the electron theory of atoms, and the still more subtle relationships of the quantum theory are examples of this in the physical sciences. The relationships which are now being determined between complex molecular arrangements of DNA and so on, and genetic codes are an example in the biological sciences. In economics we have such things as price theory and macroeconomic employment theory, among others, and, in social science, models of population dynamics, organization theory, role structures and so on, all of which involve essentially a set of relationships. Some of these relationships, perhaps, are identities, some are descriptive of empirical behavior, which are consistent with certain well-defined patterns in space-time.

Equilibrium models are among the simplest relationships. Here the relationships which operate on the state operate to perpetuate it, that is, to reproduce it in successive time periods. The great advantage of equilibrium models is that, if the equilibrium is stable, the particular dynamics of the system is relatively unimportant because, no matter what the dynamic path following a disturbance, the system will approach equilibrium again. A ball in the bottom of a bowl is a simple example. If it is disturbed, it may follow all kinds of dynamic paths around the bowl, but it will eventually come to rest in the same place where it was before. We can expand equilibrium models to include comparative statics, that is, a succession of equilibrium models, and here again the actual dynamic processes which lead to the successive equilibria may be unimportant.

There are a number of varieties of equilibrium models. One group includes pendulum-like models, in which a divergence from the equilibrium position produces dynamic forces to bring the system back to equilibrium; the strength of the forces is indicated by the extent of the disturbance. Then there are what might be called "boundary equilibria" in which a system may fluctuate freely within a certain set of states, but when it approaches a certain boundary, forces will be set in motion to reverse the direction of the system, and so prevent its crossing the boundary. Another type of equilibrium is homeostatic equilibrium, such as the thermostat, in which divergence from the equilibrium, or "ideal" position in the system, simply sets forces in motion to change the position of the system towards its equilibrium as long as there is any divergence. Most equilibrium systems, it should be noted, produce fluctuations around equilibrium or within the equilibrium set, which may be more or less regular in character, depending on the nature of the system.

Generally, dynamic systems which have no true equilibrium are much harder to handle in model building. Nevertheless they are very important in nature. It can be said, indeed, that nature knows no true equilibrium, that all equilibria in nature are temporary. The living body, for instance, may exhibit temporary homeostasis, but it is subject to irreversible processes of growth and decay and aging, which eventually lead to death. The equilibrium of erosion of land forms is constantly being disturbed by processes of mountain formation. Ecological equilibria are constantly being disturbed by mutations of many kinds. Evolution, indeed, is profoundly against the process tending torward equilibrium; it involves irreversible change. Human history is, on the whole, an evolutionary process *in the field of the human nervous system,* and is subject to much the same principles. In systems of these kinds it has to be dynamics or nothing, for where there is no true equilibrium, all that there is in the system is dynamic processes, and its history and future depends solely on these ongoing processes.

Crude notions of cause and effect are examples of primitive dynamic models in the systems of "folk knowledge." As we move into scientific systems, the model of cause and effect gives way to concepts such as mutual determination, mutation, and selection, mixed systems which are partly determined and partly random, and such concepts as "threshold" or "watershed" systems, in which some quite slight dynamic impulse may push the system over some kind of threshold into a wholly new dynamic process. This, in turn, may lead towards a position of equilibrium or a path through time quite different from what it had before.

The eighth step in the development of epistemological significance is the search for evaluative indices, that is, indicators which go "up" (or their inverse which may go "down") when the world is judged to get "better." We have to face the fact that no single indicator which can serve this purpose has been been found or is ever likely to be found. Aristotle indeed perceived this when he developed the concept of the "golden mean," which implies that the relationship between any particular indicator and the ultimate good is nonlinear and will exhibit a maximum. When any quality, property, or indicator increases from low levels this is likely to be "good." If it gets to very high levels, however, an increase is very likely to be "bad." At its "Aristotelian Mean" its "goodness" will be a maximum. What we are dealing with here is what economists have called a "welfare function" and which might perhaps be more generally named an "evaulation function," such as Equation 1, in which we suppose that there is some value of the universe, *V*, which may incidentally be an ordinal number rather than a cardinal number. This we postulate is some function *F* of

$$\overset{\text{The Universe}}{V = F(a, b, c, \ldots)} \qquad \text{Equation 1}$$

the relevant part of the universe, *a ,b, c. . . .* The arguments *a, b, c, . . .* can be anything whatever which affects our evaluation of the total system. No such function, of course, is known in any explicit form. Nevertheless, any evaluative process implies something like it, and we can discuss some of the arguments which are inside the bracket. We can also discuss the nature of their differential coefficients, that when *a* goes up, does *V* go up, or does it go down? If when *a* goes up, and *V* goes up, does it go up a lot or a little? We can also discuss, as we have seen, whether the relation is nonlinear, as it usually is, so that when *a* is small and it goes up, *V* goes up too. Then, when *a* is at its Aristotelian Mean, a change in *a* does not change *B*, and then when *a* is large, beyond its ideal value, an increase in *a* will diminish *V*.

Another very important property of the evaluation function is the interrelations of the elements within it. Suppose, for instance, that when *a* goes up, *b* goes down. Then the total effect of an increase in *a* on *V* depends on the effect of the decline in *b* on *V* also. These kinds of effects are often neglected by naive people who do not understand the subtle interrelationships of the social system. Thus, a person who uses insulting behavior because he thinks an increase in insult to those who deserve it increases the value of his self-image is often very much surprised to find that the people whom he has insulted treat him badly and either increase something which has a negative effect on his *V* or decrease something which has a positive effect on it. The decision to involve the United States in Vietnam might have been very different if the people who made it had been more aware of the effects of this decision on internal dissent in the United States and on the legitimacy of established authority.

A criticism of a particular indicator, such as the Gross National Product, as an evaluative index, may take two forms, which are rather different, although sometimes confused in practice. The criticism may relate to the effect on V, given the evaluative function of the particular indicator; that is, we may say that if a is the GNP, then an increase in a does not, in fact, increase V, perhaps, on the Aristotelian Mean principle that beyond a certain point it is bad for people to get richer. The second kind of criticism relates to the interrelationships among the variables on the evaluation function. We may think that a rise in a, that is, the GNP, may in fact correspond to a rise in V if other things remain equal, but we point out that a rise in a creates, shall we say, a rise in b, which is pollution, which operates to diminish V, or a rise in c, which is vulgarity, which also operates to diminish V, or a fall in d, which is equality, which also operates to diminish V, and so on. The sum of all these movements may diminish V, whereas we might still think that a simple rise in the GNP would raise it.

Much conflict in society arises because different individuals have different evaluation functions. As Kenneth Arrow has demonstrated,[2] there is usually no mechanical way in which the evaluation functions of different individuals can be summed into a "social welfare function" which will give a unique answer

2. Kenneth Arrow, *Social Choice and Individual Values*, 2nd ed. (New York: Wiley, 1963).

to the question whether the state of the universe *A* is better for everybody or for society as a whole than state of the universe *B*. Nevertheless, the whole political process in society is concerned with the making of decisions which involve assessments or evaluations of complex systems in the light of an evaluation function for a whole society, or at least for parts of it. There are decisions which are "made on behalf of" and these inevitably involve a social welfare function on the part of the decision-maker. A considerable amount of decision theory is—or should be—concerned with how we develop political structures, inventions, habits, constitutions an so on, which in fact make the private evaluation function of powerful decision-makers correspond to some functions of social evaluation by society as a whole, so that what is perceived by the powerful as good for them is also perceived by the rest of us as good for us. This, however, is a problem which is very far from having been solved.

Even though there is no way of defining exactly and unequivocally, in a way that will command universal agreement, a total social evaluation function, it may be possible to break the overall system of society down into subsystems in which the evaluation function can be defined with a greater degree of certainty, so that we can be a little more certain in the subsystems when things are going from bad to worse or from bad to better. This is what I have called elsewhere "the theory of deteriorating systems"—exhaustion of resources, deterioration of both physical and human capital, the increase in "negative commodities," such as pollution, the development of perverse dynamic processes, like arms races, unmanageable conflict, and the development of malevolence, failures in the transmission of knowledge, or the spread of "knowledge pollution," such as error based on superstition; all of these can be identified roughly as subsystems in which it is not too difficult to tell whether we are going "up" or "down." Most of these deteriorating systems derive from two closely related processes. One is the failure of human learning; the other is a failure of the system of "payoffs," that is, rewards and punishments. If vice is rewarded and virtue is penalized, deteriorating systems are likely to result.

Before going on to apply these principles to economic indicators, we should just mention a property of all epistemological structures which is frequently overlooked and yet is of enormous importance in the making of decisions; this is uncertainty. All images of the world, no matter at what level, have some degree of uncertainty, even if in some cases this may be negligible, as for instance in a road map, which is a model of the world in which we usually put almost 100 per cent confidence. As we go to more and more complex levels of the epistemological hierarchy, however, uncertainty tends to become greater. Tests of statistical significance, indeed, are closely related to an attempt to measure uncertainty, and apply to all levels of epistemological significance. One of the defects in the current training in the social sciences is that there is a certain tendency to confuse statistical significance with epistemological significance. These are not the same thing at all, although they are related through the concept of uncertainty. One of the real dangers of elaborate "staff" analysis, which is particularly noticeable in simulation and gaming, is that this may create an illusion of certainty for the powerful decision-maker in a situation which is objectively uncertain. This can easily lead to disastrous decisions. The type of decision which is appropriate under uncertainty may be quite different from what is appropriate under certainty. Under conditions of uncertainty, he who hesitates is saved. A high value should be placed on lack of commitment, liquidity, and reversibility of decisions. Under conditions of certainty, classical "maximizing behavior" is much more appropriate. We find what is the best thing to do and "zero in" on it. Under uncertainty, zeroing in almost always leads to hitting the target of disaster.

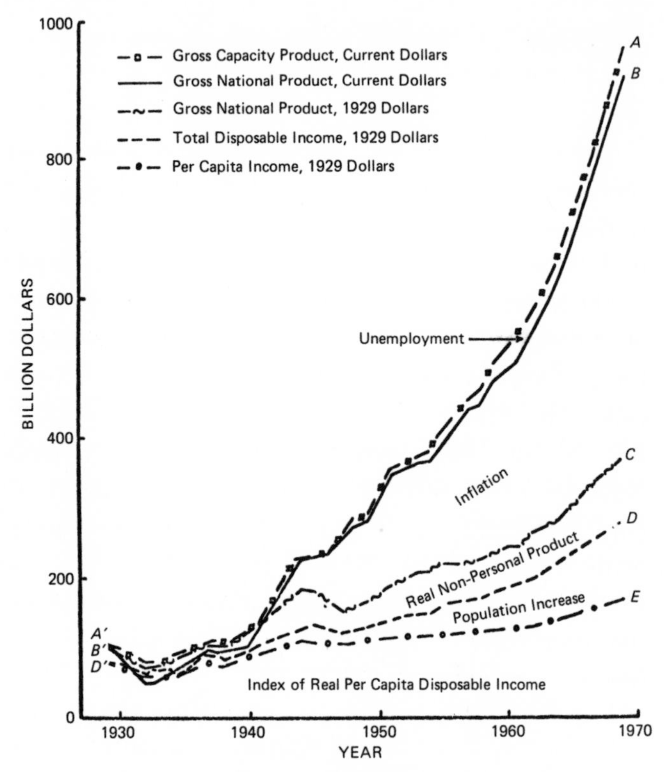

Figure 1. Components of the Gross Capacity Product, 1929–1969

II. *What Do Natural Income Statistics Really Tell Us?*

In the light of the "hierarchy of epistemological significance," let us now look at national income statistics, especially those of the United States, to see both what can be learned from them and what might be their defects. Figure 1 represents a number of indicators, both in absolute and in ratio terms for the last forty years. The line *A* is what I have called the "Gross Capacity Product," in current dollars. Ideally, this would represent what the Gross National Product would be in the absence of unemployment. A number of calculations have been made to produce this figure.[3] The simplest of these would be to expand the Gross National Product by the proportion of the labor force unemployed, r, that is,

$$\mathrm{GCP} = \frac{\mathrm{GNP}}{1 - r} \qquad \text{Equation 2}$$

This, however, may be misleading in the light of the very large changes in the "war industry," that is, the

3. Stanley W. Black and R. Robert Russell, "An Alternative Estimate of Potential GNP," *The Review of Economics and Statistics.* Vol. LI, No. 1 (February 1969), pp. 70-76.

defense sector of the GNP, so that I have deliberately chosen a somewhat modified version of this definition. I have calculated a Gross Civilian Product, which is the Gross National Product minus National Defense, and then expanded this Gross Civilian Product by the proportion of the labor force unemployed to get a Capacity Civilian Product, then added the National Defense sector to that, to get a Gross Capacity Product. Thus, if *W* is the absolute dollar size of the war industry, we have

$$\text{GCP} = \frac{\text{GNP} - W}{1 - r} + W$$

$$= \frac{\text{GNP}}{1 - r} - \frac{Wr}{1 - r} \qquad \text{Equation 3}$$

The main justification of this formula is the assumption that the size of the war industry is determined by forces in the international system quite extraneous to the econnomy and hence that Unrealized Product is all in the Civilian sector. This method of calculation perhaps somewhat underestimates the Gross Capacity Product, but its further refinement seems unlikely to produce very much in the way of results, or if it does, it is a compromise between both extreme simplicity in the first formula and what seem to me unnecessary complexities in further refinements.

Line *B* is Gross National Product, in current

dollars, and Line *C* is the Gross National Product, in 1929 dollars. The difference between *C* and *B* is, of course, the degree of inflation since 1929, and we see dramatically that since the Second World War at any rate this has been an age of inflation. There is some doubt about the significance of the price deflators, which have to be calculated on the base of certain rather arbitrary assumptions, especially over a period as long as this. They usually express changes in quality and composition of the commodity mix inadequately. This problem is inherent in the concept, however, and there is not much that we can do about it.

Line *D* represents the Total Disposable Income, in 1929 dollars, that is, income which accrues to households, after taxes. I have called the difference between *C* and *D* "Real Nonpersonal Product." It roughly includes the Capital Consumption Allowance, plus the government sector of the economy, insofar as that produces "public goods." Each of these four indicators have some significance, though it is by no means easy to estimate what they mean in terms of welfare. The line *E*, which is Per Capita Disposable Income, in 1929 dollars, which is the ratio of Total Disposable Income to Population, is perhaps the most significant figure from the point of view of average economic welfare. It can be subjected, however, to a good deal of valid criticism. It underestimates economic welfare insofar as it does not take household production into account, which is not included in the Gross National Product; also it underestimates economic welfare insofar as the price deflators overestimate the inflation, and it denies that individuals take private satisfaction from public goods. It overestimates economic welfare insofar as it does not take adequate account of deterioration of the environment, or of items of personal expenditure which should really be regarded as "costs" rather than "income," such as commuting, health maintenance, educational investment, and so on.

No serious attempt has been made to estimate any of these additions or subtractions from per capita disposable income in order to achieve a "true" index of economic welfare. The overall task indeed is probably impossible, for even "economic" welfare in the last analysis is essentially a psychological magnitude of which any mere commodity total is a very imperfect measure. Even on the commodity side, however, we ought to be able to do better than we now do. One thing that Figure 1 brings out dramatically is that the enormous increase in Gross National Product or Gross Capacity Product, in current dollars, immensely exaggerates the increase in welfare (as measured, however imperfectly, by line *E*). If we were to deduct pollution, education, health, and commuting, as costs, from the Per Capita Disposable Income, the rise in the last forty years might look very modest indeed. It is very doubtful whether Americans are as much as twice as rich as their grandfathers, which is clearly a modest rate of real economic growth.

Figure 2 shows the proportions of the Gross Capacity Product as broken down among its major constituents, measuring from the bottom up—Personal Consumption Expenditure, Gross Private Domestic Investment, Net Exports, State and Government Purchases of Goods and Services, National Defense (which, of course, is a Federal Government Purchase), Federal Government Purchases Excluding National Defense (that is, Federal civilian government), and finally, the Unrealized Product. A great deal can be learned from these proportions. We see, for instance, that the first twenty years of this period was an era of tremendous instability, whereas the last twenty years have been a period of quite surprising proportional stabilities. The Great Depression is dramatically shown as Unrealized Product, which is roughly equivalent to Unemployment; it rises to nearly 25 percent at *N*. We see the relative failure of the New Deal, from *N* to *P*, the enormously traumatic depression of 1938 at *Q*, and then the virtual abolition of Unemployment from *Q* to *R*, with the increase in National Defense expenditures during the Second World War, which goes from less than 1 percent of the economy in the 1930s, to about 42 percent in 1944. We see, also, dramatically, how the Great Depression was associated with a decline in Gross Private Domestic Investment almost to zero by 1932. We see also

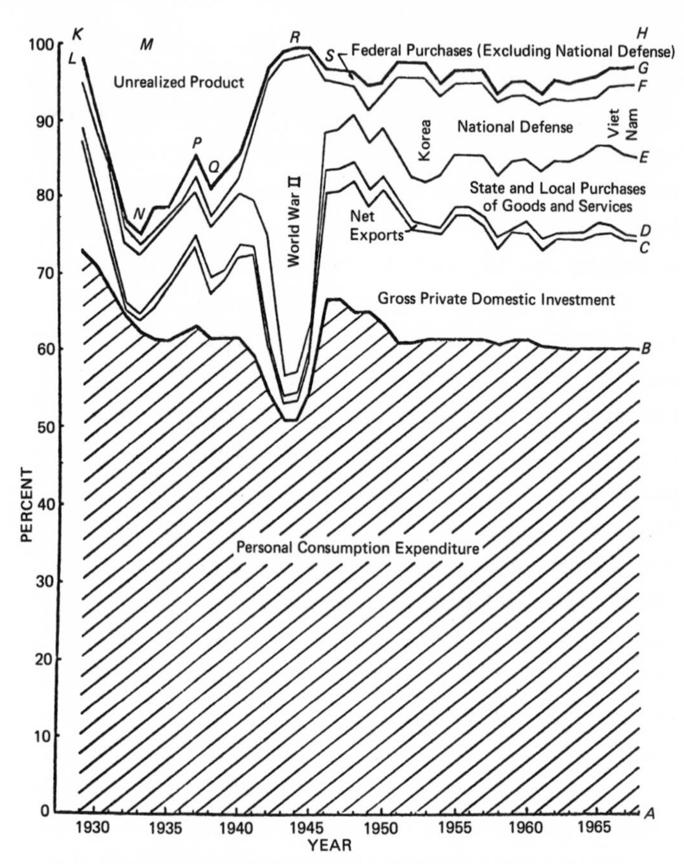

Figure 2. Unrealized Product and Major Components of GNP as a Per Cent of Gross Capacity Product

how the depression of 1938 is associated with another collapse of Gross Private Domestic Investment, and we see how the Second World War was "financed" in real terms, about half out of Unemployment and about half from a reduction in the other items, particularly in Gross Private Domestic Investment and in state and local government.

We see the impact of the Marshall Plan in the expansion of Net Exports in the years between 1946 and 1948, but we also see that normally Net Exports have been a very small part of the American economy, which is overwhelmingly domestic. Net Exports, rather, underestimates what might be called "American economic imperialism," but even at that it is clearly small potatoes from the point of view of the total economy. The relative stability of Total Government Expenditures in the last twenty years as a proportion of the economy is also very striking, as is also the relatively small magnitude of the Vietnam War, even by comparison with the Korean War, on the economy. This very fact should warn us to be careful in the qualitative interpretation of these quantitative phenomena, for the impact of the Vietnam War on American society has been very much larger qualitatively than was that of the Korean War, in spite of the fact that quantitatively it is not as large a proportion of the economy. The little business cycle of the 1950s is shown rather clearly and is revealed also as originating mainly in Gross Private Domestic Investment, and a further breakdown would show that it originated mainly in an inventory cycle. The absence of the business cycle in the 1960s also is very striking, though by 1970 we managed to create one again.

Figure 3 now shows another very instructive breakdown, this time of National Income in terms of its components—Compensation to Employees, Business and Professional Income, Income of Farm Proprietors, Rental Income of Persons, Corporate Profits plus Inventory Valuation Adjustment, and Net Interest. The two components of Gross Capacity Product which have been mainly responsible for disturbance—the Unrealized Product and National Defense—are shown for comparative purposes above the line. The impact of the Great Depression of the 1930s is particularly striking. We see how Corporate Profits were eliminated altogether and became losses in 1932 and 1933, squeezed out by the rise in Net Interest as a result of the deflation, and by the rise in the proportion of the national income going to labor. We see a surprisingly little disturbance in the proportions of the national income as a result of the Second World War. There is a certain squeeze on Profits, perhaps as a result of price and wage control. In the last twenty years, again, the proportions of national income have been astonishingly stable. We see a decline in agriculture. There seems to be a very slight rising trend in Compensation of Employees, but apart from this the proportions are remarkably constant.

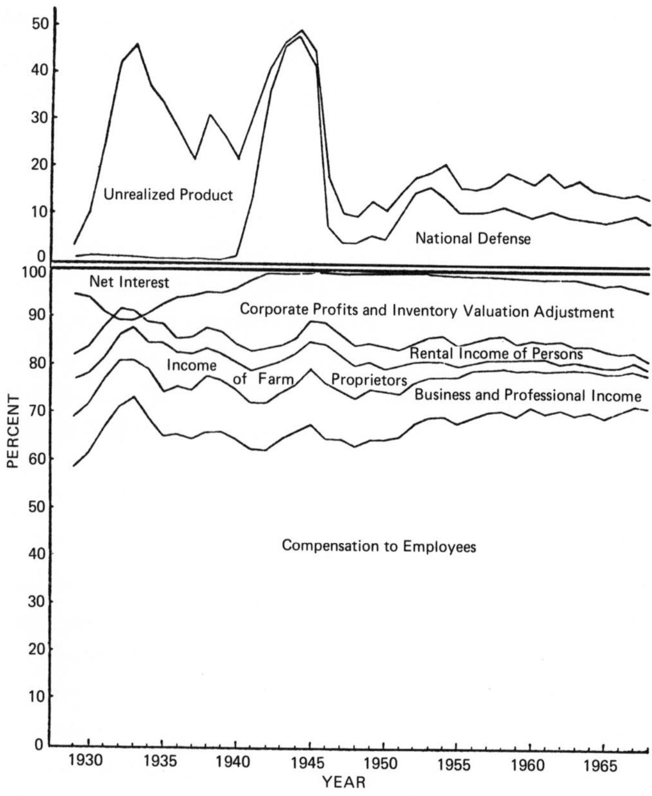

Figure 3. Components of National Income, and National Defense and Unrealized Product as a Per Cent of National Income

Even as it is, Figure 3 is very instructive, but both the aggregate which it represents and the breakdown is very far from being satisfactory. The national income itself is a very odd concept in the national accounts. It is roughly the net of indirect taxes and the gross of direct taxes. Thus, it is neither a measure of income before taxes nor of income after taxes, but something in between, and there seems to be very little excuse for this particular concept, except for certain computational convenience. What we would really like to have would be a breakdown of something like the Net National Product, which is fairly close to National Income, plus Direct Taxes, that is, income before taxes, and a breakdown of Personal Disposable Income, that is, income after taxes. The only justification I can think of for the present national income concept would be to assume that somehow direct taxes were paid for something, and hence represented public goods which could be clearly allocated to households, whereas indirect taxes are not paid for anything very much, except for the overhead of government. The very absurdity of this justification, however, underlines the fundamental absurdity of the concept itself. It is, however, the only major concept of national income for which clear breakdowns can be obtained, and though the general conclusions would probably not be changed much if we had a better aggregate concept, it is very unsatisfying, to say the least, to have to work within the existing framework.

It is not only the aggregate that is unsatisfactory; the breakdown itself as given in the National Income Accounts is extremely arbitrary and really mixes two different possible characterizations. The Income of Farm Proprietors, for instance, would better be an item in a breakdown by industrial groupings. The Rental Income of Persons is a kind of residual with very little economic significance. It represents partly that proportion of the real estate market which is not incorporated, and also a certain amount of private income from royalties, but it is certainly not significant as a segment of the economy. The inadequacy of the existing breakdowns is shown by the fact that it is virtually impossible to allocate the national income of the United States or any other similar aggregate, such as that between labor income and nonlabor income, for the items of Business and Professional Income, and Income of Farm Proprietors are a mixture of labor and nonlabor income. It is not even possible to separate Business from Professional Income, and the procedures by which this item is estimated suggest that it may be much less accurate than some other items. It is probably the main source of the "statistical discrepancy" in the accounts. Another weakness of the accounts is that they do not show clearly the distribution of income, as between the Corporate and the Noncorporate sectors of the economy, though the sum of the Compensation to Employees and Corporate Profits and Net Interest certainly between them account for most of the Corporate sector. Figure 3 suggests that there has been a noticeable, though not large, corporatization of the American economy in the last forty years. Much of this, however, is accounted for by the decline in the proportion of the economy that is based on agriculture.

Figure 3, with all its defects, does bring out some highly characteristic properties of the American economy. One is that it dispels the illusion that war is "good for business," for it is clear that both in the Second World War and in the Korean War, Corporate Profits were somewhat squeezed; otherwise fluctuations in the National Defense do not show up very much in the distribution of income. The figure also points out that the rise of the labor movement had very little effect on the distribution of income. If indeed we take the period from 1933 to 1943, which saw the great rise of the labor movement, from a little over three million members to over fifteen million members, and an even larger rise in the proportion of American industry brought under collective bargaining, we see that in this ten years the proportion of national income going to labor actually feel quite sharply and that fluctuations in this proportion have been due much more to influences outside the labor market, such as the Great Depression or the Second World War, unless we suppose that the slight long-term trend in the rise in the proportion of national income going to employees is a result of labor organization. Which ever way the impact lies, it is clearly quite small.

TABLE I

The Depression of 1970
Selected Items from the National Income Accounts,
Expressed as a Proportion of the
Gross Capacity Product,
1969 and 1970

	1969	1970	Change
Gross Capacity Product, absolute current dollars	962.3	1023.6	
1. Percent of Gross Capacity Product	100	100	
2. Gross National Product	96.8	95.5	−1.3
3. Unrealized Product	3.2	4.5	+1.3
4. Personal Consumption Expenditures	60.0	60.3	+0.3
5. Net Exports	0.2	0.3	+0.1
6. Government Purchases, Federal Civilian	2.3	2.3	0
7. Government Purchases, Federal Defense	8.2	7.5	−0.7
8. Government Purchases, State and Local	11.5	11.8	+0.3
9. Gross Private Domestic Investment	14.5	13.3	−1.2
10. Change in business inventories	0.8	0.3	−0.5
11. Fixed investment	13.7	13.0	−0.7

12. Residential Structures	3.3	2.9	−0.4
13. Nonresidential	10.3	10.0	−0.3
14. Nonresidential structures	3.5	3.4	−0.1
15. Nonresidential Producers' durables	6.8	6.6	−0.2

As a final exercise in what might be called "proportions analysis" of the economy, Table I illustrates the depression of 1970. The Unrealized Product, as we see in line 3, rose from 3.2 percent of the Gross Capacity Product to 4.5 percent, a change of 1.3 percentage points. These have to be accounted for in the rise or fall of other elements of the Gross Capacity Product. We see that there were declines in Defense and in Gross Private Domestic Investment, amounting to 1.9 percentage points (line 7 plus line 9) and an offsetting rise in Personal Consumption, Net Exports, and State and Local Government Purchases of 0.7 (lines 4, 5, and 8). The total is not quite 1.3 because of rounding. Of the 1.2 percentage point decline in Gross Private Domestic Investment, we see this is divided between a decline of 0.5 percentage points in Business Inventories, and 0.7 in Fixed Investment (lines 10 and 11). Of the Fixed Investment, 0.4 percentage points was accounted for by Residential Structures (line 12), 0.1 by Nonresidential Structures (line 14) and 0.2 by Producers' Durables (line 15). Looking at the overall picture, therefore, we can say the decline in Defense and in Inventories and Fixed Investments was not sufficiently offset by a rise in Civilian Government, especially State and Local, and that this was the basic cause of the depression. If we contrast this experience, incidentally, with the experience of 1963 to 1965, there was an even greater decline in the proportion of the Gross Capacity Product in Defense, but this was more than offset by a rise in other factors, especially in State and Local Government, so that the overall unemployment declined. If we ask ourselves what factors could have been different in 1970, which would have given us no depression at all, the answer is again very clear. If state and local governments had not been so stingy and had expanded their purchases, or if the purchases in Structures had not declined so sharply, no doubt as a result of high interest rates, the depression would have been avoided. The example does illustrate the problem, however, that avoiding depression requires offsetting changes in a considerable number of different items, movements in which may be unrelated. A genuinely cybernetic apparatus in the economy would see to it that a decline in one item would set forces into play to produce increases in other items. This is something that up to now we do not really have for small depressions, although we probably do have something like this for larger ones. Certainly if Unemployment goes to 6 percent, not only do political alarm bells ring all over the place, which produce offsetting increases in government expenditure, but this very activity on the part of government probably encourages the private sector to expand likewise.

At the next level of analysis, syndromes, the main statistical technique is correlation and regression analysis, though this by no means exhausts the structure of possible patterns. In this brief essay we can do little more than mention this vast field of research and to warn against the dangers of purely mechanical statistical analysis. The more complex the syndrome, the more important it is to recognize that statistical significance is not the same as epistemological significance.

A single example will have to suffice to illustrate the problem. Figure 4 illustrates some of the problems in perceiving syndromes on the rather larger scale. Here we plot the per capita GNP against the rate of growth of GNP per capita for a considerable number of countries for the period between 1950 and 1960. A simple correlation would lead us to conclude that there was very little relationship between these two variables, and what there was was positive, in the sense that the richer the country the higher its rate of growth. If, however, we divide these countries into two groups, which I have called the "A" countries and the "B" countries, certain relationships are at least suggested, though much better longitudinal studies than we now have would be necessary in order to confirm this. For the "A" countries, which are mostly in the Temperate Zone,

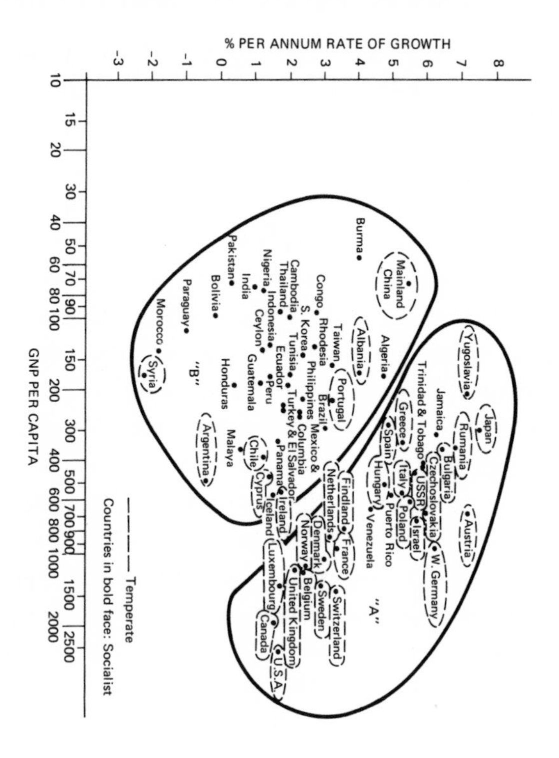

Figure 4. Per Cent Per Annum Rate of Growth, 1950–1960

the syndrome which is suggested is that of a negative relationship between per capita GNP and rate of growth, suggesting that the richer the country the slower it grows. This, indeed, is what we would normally expect, for virtually every growth process eventually runs into diminishing rates of growth. The relationship would suggest, furthermore, a zero rate of growth somewhere between 5000 and 10,000 dollars per capita at 1958 prices. The "B" countries, in contrast, exhibit very little relationship between rate of growth per capita GNP, suggesting that they are participating in a much more random process. I use this as an example of the difficulty of perceiving syndromes, especially through time, and the advisability also of being extremely cautious in the interpretation of correlations. This is another case where statistical significance may not be the same thing at all as epistemological significance. Relationships of any kind, particularly correlational relationships, are significant epistemologically only insofar as they operate to test models. In themselves they may have very little significance, except in so far as they point towards models which need to be tested.

The next stage of epistemological analysis, which is model building, is perhaps the most significant because without models we cannot hope to make predictions, even conditional predictions, and without conditional predictions, models cannot be tested, and knowledge cannot really grow. Some models are extremely simple, such as the proposal that if I go to such and such a place, I will find the post office. These are fairly easy to test. All knowledge, including folk knowledge, involves the formation and testing of models. When we get to complex systems, especially macrosystems, both the formation of models and their testing becomes increasingly difficult, and there may be a wide gap between statistical significance and epistemological significance. We will have to content ourselves in this brief paper with a single illustration. During the Great Depression of the 1930s, President Hoover really had no model of what was going on at all, and hence was quite incapable of dealing with it. President Roosevelt, indeed, was not much better off, as we see from the relative failure of the New Deal and from the fact that whatever recovery there was in the 1930s was mainly due to the recovery of Gross Private Domestic Investment, which had very little to do with New Deal policies. The Keynesian equilibrium model throws some light on the problem, especially when expressed in terms of the proportionalities of Figure 2. What we see happening from 1929 to 1932 is an enormous decline in Gross Private Domestic Investment, almost indeed to zero, with an associated decline in Consumption, as a result of declining incomes, and without any compensating increase in Government Purchases. We can simplify the model by neglecting Net Exports and writing Y=Income, C=Consumption, I=Investment, and G=Government Purchases. Then we have the basic identity:

$$Y = C + I + G \qquad \text{Equation 4}$$

Then let us suppose a very simple consumption function,

$$C = mY \qquad \text{Equation 5}$$

and then we have

$$Y = \frac{I + G}{1 - m} \qquad \text{Equation 6}$$

so that if m and G are constant, Y will diminish as I diminishes.

This model, however, does not explain why Investment declined so severely. In order to explain this, we have to have a further model which suggests that Profits depend on Investment, and Investment in its turn depends on Profits. This is suggested by comparing Figures 2 and 3. Here it is clear that a decline in Profits and a decline in Investments are closely correlated. We have to have a model, however, and a dynamic model at that, if we are to understand which operates on which. In order to get a model, however, we almost have to postulate some identity, for the mere observation of a relationship rarely gives us a satisfactory model. The simplest identity here is that Profits are equal to Business Savings minus Business Distributions, and Business Savings are roughly equal to Investment, in current dollars, plus an increase in Business Stocks of Money. Decline in Investment, therefore, which is not offset by an increase in Business Distributions in Dividends and Interest, will create a decline in Profits. The decline in Profits, however, if it produces expectations of further decline, is likely to result in a decrease in Investment, which produces a further decline in Profits, and a further decrease in Investments, and so on, until we slide all the way down to 1932.

The very interesting question of tremendous practical importance is whether a process like this could happen again, for it has not happened in the last thirty years. The answer is perhaps, that under some circumstances, it could. While the American economy now has good cybernetic machinery for dealing with any rise in unemployment due to a decline in Consumption or even a decline in Government Expenditures, there is virtually no apparatus for dealing with a situation of a sharp decline in Gross Private Domestic Investment, which would destroy profits, and hence easily lead to the kind of slide which occurred between 1929 and 1932. This is indeed an alarming conclusion, and it should encourage us to give much thought to possible defenses against another great depression, one very important element in which it would have to be an undistributed profit tax, in spite of the fact that our experience with this in the 1930s was not a very happy one.

Another question which we must pass over all too briefly is the main unsolved problem of what might be called "governed market societies," which is how to achieve full employment without inflation. The economics profession has virtually dismissed this problem as insoluble; we certainly seem to be giving up on it too soon. An adequate model here, however, would require much more knowledge than we now have about the actual dynamics of price and wage increases. In one sense it may be that National Income Statistics have proved a handicap here because they have diverted attention from the essentially dynamic problems of the microstructure. Thus, the study of the influence of information systems or fashion in the raising of prices and money wages, requires perhaps more of an epidemiological type of model than the ones economists now use. It may be indeed that the very success of National Income Statistics in only part of the field has been disastrous from the point of view of some wider questions of economic policy. One of my revised proverbs indeed is that "Nothing fails like success," because we do not learn anything from it, and the present position of economics may very well be a disastrous illustration of this principle.

Finally, we come to the last and most difficult problem of all, which is that of the evaluation of economic systems. The problem of general evaluative functions, as we have seen, is very difficult indeed, though progress can certainly be made towards better analyses of these problems. The Gross National Product has been rightly criticized as a simple evaluative index (particularly in the brilliant article by Sametz[4]). We could certainly improve National Income Statistics substantially by taking explicit account of household production, which would not be too difficult, by including "bads," that is, negative commodities, in the Gross National Product as a negative item, and by separating out, better than we do now, public goods from private goods. When we have done all this, however, we will still not have answered all the critics, especially those who wish to impose their private values on the social system, like Mr. Berle.[5] Economists can certainly be criticized for assuming that individual personal preferences are sacred and may not be questioned. We are indeed still tarred by the Benthamite brush and, if somebody likes pushpin better than poetry, we do not generally question this taste. This does, however, neglect the problem, as Veblen pointed out, that tastes are learned, and hence cannot be taken for granted. To the contrary, there is a becoming modesty in the economist's hesitation to criticize tastes, which contrasts to my mind rather

4. A. W. Sametz, "Production of Goods and Services: The Measurement of Economic Growth." Sheldon and Moore (eds.), *Indicators of Social Change* (Russell Sage Foundation, 1968).

5. A. A. Berle, "What GNP Doesn't Tell us," *Saturday Review.* Vol. 51 (August 31, 1968), pp. 10-12.

favorably with the immodest assumption of many critics that they personally know what is good for everybody, and hence individual tastes are not important. This indeed is the major moral weakness of most radical positions. In the name of loving mankind, radical critics actually despise the tastes of the vulgar majority and wish to impose their own, of course more refined, tastes on mankind as a whole.

It is hard to take a firm stand on either extreme here. However, it is also very difficult to find a workable Aristotelian Mean, between the taste-neutrality of most economists and the missionary zeal of the moralists. If my personal view inclines more towards the economists than towards the moralists, this comes out of putting a high value on modesty in political aspirations, a value which is certainly not shared by everybody. Evaluation functions do in fact develop in a kind of ecological-evolutionary process of competitive rhetoric. We really know very little about this process, and I suspect, uneasily, that more knowledge, if it were available, might be very dangerous. Still there is no way back to the Eden of unself-consciousness and we will have to be prepared in the future to evaluate self-conscious evaluation itself.

ANTIPOLLUTION POLICIES, THEIR NATURE AND THEIR IMPACT ON CORPORATE PROFITS

Elvis J. Stahr

I. *Our Deteriorating Environment*

AT THE OUTSET let it be said that I am not an economist. But a very good one has collaborated in this paper, Dr. Robert K. Davis, who is now Resource Economist for the National Audubon Society.

The "Quote of the Day" in the *New York Times* one morning might be quoted to introduce the general theme of this chapter: "May there only be peaceful and cheerful Earth Days to come for our beautiful spaceship Earth as it continues to spin and circle in frigid space with its warm and fragile cargo of animate life." (This was said in Central Park by United Nations Secretary General U Thant during a celebration of a special Earth Day, the first day of spring!)

Most of our daily concerns seem superficial and

transitory in the perspective of the hundreds of millions of years during which there evolved on this little planet a system which makes *life* possible, the system we sometimes call the balance of nature, or the web of life, or our natural environment, or the ecosystem. The first rocks which the astronauts brought back from the moon appeared to be about 4.6 billion years old. In all that time life as we know it, if any at all, has not emerged on the moon. Yet it has developed on Earth in rich variety, through the functioning of an ever-evolving system of interrelated cycles and forces. It is a highly intricate system, or process, based on complicated and continuous interactions of air, water, soil, sunlight, and countless species of plant and animal organisms. Not one is independent of the others. This life support system, unique in the universe so far as we know for certain, and in any case the only one *we* have, is now in trouble, even perhaps in danger, because of the activities of one species, the one called man, ironically the only one which is supposed to have intelligence and foresight.

All of us presumably know the three vital "fives": Man can survive five weeks without food, five days without water, five minutes without air. Yet we seriously pollute the air and water of our ecosystem, and we are now reducing its capacity to produce food. The situation is at last beginning to be recognized, and today nearly everybody has heard something of it. Many are calling it "the crisis of the environment," and there is even to be a world conference on it, in Stockholm in June, 1972, called by the United Nations.

Last November, I attended a conference in England on the subject of pollution control, a conference of some thirty government, industry and university people, plus a couple of conservationists, from the United Kingdom, the United States, and Canada. Though some of us were more alarmed than others, all agreed that the environmental situation is serious. I might explain that it is also relatively new, resulting as it does mainly from the impact of two coinciding and enormously powerful forces unleashed by man since World War II: the "explosion of population" and the "explosion of technology." Between them, these two forces were having by 1971 a perceptible impact, largely adverse, on the life-support system on Earth—and projections of their continuing growth at comparable rates could cause even conservative conservationists to foresee consequences of great gravity. How did this happen?

For one thing, we have used our highly developed technology on a gigantic scale—and often brilliantly, let it be acknowledged—to solve narrowly focused, relatively short-range problems in such fields as transportation, power generation, flood control, industrial production, and amenities and services for a fast-growing population *but* with grossly inadequate attention to the far-reaching effects of the resulting environmental alterations. In fact, we ignored some of the simplest basics of man's relation to nature for a long time. Planning (if any) at every level, from the private "developer" to the national government, too rarely took ecological factors seriously into account. The ancient injunction that man should go forth and multiply and have dominion over the Earth and all that dwell therein seems often to have been interpreted as a kind of mandate not only to overpopulate but to alter the Earth at will—and thus to risk destroying all that dwell therein—rather than as a legacy of Earth's resources in *trust* for man to use wisely and with restraint and a sense of the future. Man finally achieved the ability to make drastic and rapid alterations in basic processes, and in major components, of the ecosystem which nature took long periods of time to develop, but man had not and has not yet achieved the wisdom to use that ability with great care.

Thus the estuaries, nurseries of the sea, are dredged and filled for real estate developments even while population pressures point to great future needs for more food from the sea. Dams are built for the presumed protection of flood plains from occasional floods, but often many acres of the dwindling numbers of natural river gorges and forests, and even of tillable land, are permanently flooded. Airports are often sited on foundations produced by the dredging and filling of marshlands, lake shores, or coastlines, which are vital to natural life cycles. Mounting masses of solid wastes and sewage are often dumped into the nearest lake, river, or bay. Factories producing great quantities of pollutants are very often located on streams into which the effluvient is simply unloaded. Farmers, homeowners, and even some managers of public lands apply persistent chlorinated hydrocarbon pesticides, such as DDT and dieldrin, in an effort to control certain insects (which usually can be controlled by other, less dangerous means) and as a result are going to pollute lakes, streams, and even the oceans for years with substances that may inhibit the oceanic oxygen-producing organisms. They are threatening with extinction numerous wild animal species (including falcons, hawks, pelicans—even our own national bird, the bald eagle). The full effects of their application on human health may not be known for some time yet. Automobiles are built and sold by the millions with at least twice as much horsepower as most people really need and at least twice as much polluting power as city air can readily cleanse. Detergents and fertilizers by the ton find their way into waters where their nitrates and phosphates stimulate eutrophication—as in Lake Erie. And there are also problems with mercury and some other heavy metals. But that is not all. Thousands of acres of land are *paved,* in this country alone, every day, inevitably affecting underground water tables and reducing the ability of land organisms to produce oxygen. Not only carbon dioxide and monoxide but lead, rubber, asbestos, radioactive particles, toxic compounds of sulfur and nitrogen, and much else are constantly spewed into the *atmosphere,* some of them in alarming quantities. It has further been proposed to spend billions on a supersonic airplane that could pollute the *stratosphere.* Speaking of super, take super-highways. Who lays them out? No ecologist, you may be confident. There's one right now under contract, by the way, to cut through the biggest and most beautiful park in Memphis, Tennessee. The National Audubon Society, joined by others, sued to stop it. The U.S. Supreme Court recently agreed with us that the whole project was planned, and approved by two successive U.S. Secretaries of Transportation, without serious consideration of environmental factors!

To this list, add such matters as oil tanker crack-ups and underwater drilling leaks, poorly regulated strip mining, wanton killing of nongame animals and birds, littering of everything from city streets to rural picnic grounds, urban crowding, and rising crescendoes of noise in our cities, and one must conclude that ecological ignorance and apathy in contemporary society—here and abroad—are or have been almost overwhelming.

No community, state, or country is wholly free of these abuses. None can solve them all, by itself. But the United States is responsible for more of these abuses than any other nation and thus has the greater responsibility to overcome them. Nature does not know or really care where national or local political boundaries are laid out by man. We all share the same atmosphere, the same water cycles, much of the same wildlife. When any nation poisons the air or water, wastes natural resources, gets too selfish about, say, whaling, tears at the fabric of the web of life, it can produce effects felt sooner or later around the world.

All of us, as individuals, are polluters. In fact, it is the demands of *people* that have built up most of the pressures on the ecosystem. There cannot very well be any production or consumption without some pollution. It is when things get out of hand, or threaten to, that prudence and the survival instinct combine to send up warning signals.

Clearly, today, there are serious needs both for educating people about the environment and for beginning to conserve it. Most schools and colleges have done a demonstrably miserable job of educating people about the fundamentals of their relationship to nature. Ecology has scarcely been heard of, much less studied, by college students until last year, if even now. Political agencies at all levels of government until quite recently failed even to *try* to protect adequately the public's vital interest in a decent environment. At least since World War II, people themselves, almost everywhere, have been increasing their demands for energy and for consumer goods, and reproducing their species, at rates obviously greater than their little planet can indefinitely sustain. Many of them resent being urged to slow down, to stabilize some of the growth curves. And I fear that nations still are about as prone to war upon each other as to recognize that the human race shares problems big enough to *require* a common front in a *common* cause. It is almost as if a crew of sailors fell to fighting on the deck of their ship and would not stop even when the ship sprang a leak!

At this point I can almost hear you thinking, "This man *sounds* like a conservationist—long on criticism, long on opposition, short on solutions and alternatives." But, I may seem to be critical only because I have not finished yet! Our approach in the conservation movement is *not* limited to sounding alarms, awakening concerns, and fighting defensive battles, even though, regretably, all of this continues to be necessary, and much too often. We do have constructive ideas. We seldom oppose anything blindly, whatever some may say. We do, day in and day out, insist on the use of *foresight*. There are better and worse ways of doing nearly anything; we try to insist that *alternatives* be studied before irrevocable choices are made. The fact is, we conservationists are basically very conservative.

We abhor waste; we fight boondoggling; we oppose charging ahead on expensive projects the consequences of which have not been carefully studied and considered. True conservatism and true conservation are on common ground, but there is all too little conservatism or conservation in a society which tends to equate more and more with better and better.

"Ecology"—one has heard the word a lot lately, though only lately. What is it? I suggest that it's the study of the *relationships* of the many parts of nature's system to each other. But education has failed to teach us about ecology successfully because education began an almost frenzied specialization, with a concomitant loss of interest in more general relationships. Even the environmental sciences are usually studied piecemeal; nature is broken into little pieces—geology, botany, astronomy, zoology, bacteriology, ichthyology, climatology, physics, chemistry, biophysics, biochemistry, genetics, soil dynamics, and so on. But scarcely anywhere in academe are the pieces put back *together*. To me, ecology is somewhat analogous to what is called "systems engineering." By this is meant that it is important not only that each component be in good shape, that it test out, but also that all the components fit *together* and work together so that the *system* will work.

If a green, would-be astronaut sneaked aboard a spacecraft and looked around at the array of switches and buttons and lights and wires and levers and tanks and picked one or another and said, "I don't know what that's for! Man is surely more important than gadgets. I can't see what good that thing does—so I'll just throw it out and make a little more comfort for the crew in here," you'd all say, "That man is crazy." So would I. Yet when conservationists cry that we passengers and crew of Spaceship Earth should not throw out of *our* space craft, or tear up, or disconnect, particular components of Earth's life-support system before carefully trying to study their role and predict the consequences of their loss or degradation, some people try to call *us* crazy!

We all need to be very careful where we place the *burden of proof* when not all the answers are immediately apparent. Until recently, that burden has often been misplaced, I submit, and we are paying some enormous prices for that.

Actually, then, our basic position is that when you do not know, be careful; be cautious; try to find out *before* you make the leap. Let's quit learning so many lessons the hard way. Let us ask more questions, and insist on more answers, *before* tampering drastically with the ecosystem.

It is true, though, that on some issues the conservationist does take stands that involve more than just asking for more study. When there is already a hole in a pipe, we think it is better to fix it than just to study it. And, if the choice comes to our attention ahead of time, which nowadays is beginning to happen, we are likely to oppose whatever it is that we know is going to rupture the pipe. Is that radical? To me it is quite conservative.

One of the key contributions which organizations of private citizens, such as National Audubon, are trying to make toward the solution of problems is to help educate the public about ecology, of course. And we are beginning to make a little headway. But another is to help educate the public about economics—the economics of production and pollution. I get pretty tired of hearing, "Make industry quit polluting," and "Make industry pay the cost of eliminating pollution and cleaning up"—*as if* industry had some independent source of money unrelated to that by which people pay who buy industry's products. The costs of polluted air and water are borne by all of us, one way or another. I think the fairest way to clean up is to have the cost borne by the consumers of the products which—or the production processes of which—pollute. Difficulty, of course, arises when a particular company says, "We want to clean up, but we can't afford to raise our prices while our competitors go merrily along with the lower costs and therefore lower prices they'll be able to charge if they don't clean up too." The way out of that difficulty, which thus far seems more or less generally accepted, is to have standards set by law which will protect the good producers, so to speak, by compelling the bad producers also to clean up, and thus to make more likely the protection of us all. Some protest, "All this will increase the costs and prices of many things." One response is to ask, "What will it cost *not* to halt or substantially diminish pollution *itself!* If we ask *that,* the cost of a cleaner environment will look more attractive, even economically, and certainly if we place any value on considerations of public health, the aesthetic quality of our surroundings, and above all the operating condition of our life support system.

What follows, then, will be devoted to some questions of economic tactics.

What approaches, what policies, are likely to achieve major reductions in pollution without necessarily requiring basic changes in political and economic institutions? As an aside on the sometimes debated question whether the free-enterprise system and a decent environment are mutually exclusive, I know of no totalitarian or socialist system for which the record suggests that it automatically eliminates or even lessens industrial pollution. Be that as it may, a free society and a private-enterprise system can do the job, I submit, if they *want* to. If they believe or pretend that there is no job to be done, then, of course, we may indeed despair. But many industrialists and even more politicians are now acting (and more are at last talking) as if they understand the problem and want to help solve it. What, then, is the pollution problem (besides being *one* of our *environmental* problems)—and how can we account for the concern which industry and government are directing to it?

II. *The Nature of the Pollution Problem*

The American people have discovered that their atmosphere, their hydrosphere, and their land surface are being used as garbage dumps. Their growing outrage at this state of affairs is manifest in actions in their courts and by their legislatures, by federal, state and local regulatory agencies and by organizations of private citizens. Ecology, a term long reserved for use by an unheralded few, has suddenly become the by-word of a movement which has swept politicians into office and is causing shudders in the corporate world.

One may ask why the sudden concern with environmental pollution when scientists tell us that the haze of the Great Smoky Mountains is itself a form of natural air pollution and remind us that a large part of the particulate matter in the earth's atmosphere is the result of natural volcanic action. Natural pollution has been with us for all of our history. Nor does one have to search far for evidence that cities of past times suffered "accumulated horrors of their long neglect by the dirtiest old spots in the dirtiest old towns." [1]

1. Quoted by Kneese from *The Public Health as a Public Question: First Report of the Metropolitan Sanitary Association, Address of Charles Dickens, Esq. London, 1850.*

We must recognize that the sudden general concern is not freakish but is a reaction to what we may call "ecosystem overload." Man is suffering from excessive pollution in many parts of the world. Whenever natural processes cannot recycle all of the waste substances dumped into the environment, we experience the symptoms of ecosystem overload. In water, the signs are oxygen deficits, eutrophication, and direct toxicity, all leading to mortality within the aquatic community. In the atmosphere, the pollution is excessive when air currents and periodic rains are no longer adequate to keep the air cleansed of particulate matter or to dilute gases.

Since air and water are two important components of Earth's life-support system, it is not surprising that, biologically, all such pollution is damaging. We have deceived ourselves by overlooking that death is a cumulative process, not a discrete event, and that speeding up this process is tantamount to manslaughter. We are apt to ask, when we hear that an organism has died, "When did it die?" A more useful question is, "When did it commence to die?" Pollution speeds up the death process by imposing sociophysiological stress on most organisms in the polluted environment. There is abundant epidemiological evidence to show that environmental degradation increases the incidence of disease (morbidity) in man and experimental animals. Investigation of air pollution in cities has shown strong correlations between air pollution and mortality due to certain respiratory and cardiovascular diseases.[2]

A major clue to the gravity of America's environmental deterioration in the last few decades is the longevity curve of the population of the United States; for the first time in over a century, I am told, this curve has begun dipping downward although little published evidence is as yet available. There is, unfortunately, constant defensive obfuscation of such generalized evidence by the playing up of extenuating factors ("there are more poor blacks in the cities," and so on). But this will not excuse us, since ecologists and sociologists are rediscovering all over again that environmental poverty, like economic poverty, affects all of us more than we know because man is inescapably a member of the total community.

The phenomenon of ecosystem overload and our sensitivity to the set of problems which it represents deserves further explanation. Dr. Allan Kneese, in a paper prepared for the Atlantic Council, has summarized the major reasons why environmental pollution has become the number one problem in many people's minds.[3] I will recapitulate his reasons here:

2. Lester Lave and Eugene Seskin, "Air Pollution and Human Health," *Science* (1970), pp. 169, 723-732.

3. Allen V. Kneese, *The Economics of Environmental Pollution in the United States,* Washington: prepared for the Atlantic Council (December, 1970).

First: There has been an immense increase in industrial production, energy conversion, and the associated flow of materials and energy from concentrated states in nature to degraded and diluted states in the environment, which have begun to alter the physical, chemical, and biological quality of the atmosphere and hydrosphere on a truly massive scale. Furthermore, we now have the means to detect even small changes in these natural systems, and can thereby identify causes of morbidity of which we were unaware a few years ago.

Second: "Exotic" materials are being inserted into the environment. The near alchemy of modern chemistry and physics is subjecting the world's biological systems to strange inputs to which they cannot adapt quickly enough, and what adaptation occurs is limited to a few species and is therefore disruptive.

Third: Ordinary people have come to expect standards of cleanliness, safety, and healthfulness in their surroundings that in earlier times were the exclusive province of the wellborn or the rich. This I count as progress.

Fourth: We have become acutely aware of a tremendous increase in world population over the last few centuries. Doubts about how this increased population can be sustained in a finite world have crept into almost everyone's mind. A neo-Malthusianism has arisen. Rightly or wrongly, many people have come to feel that further applications of technology to force increased production of food and other goods is merely a holding operation against disaster, and that the longer we put off real solutions the more difficult it will become to cope at all.

As Dr. Kneese points out, with industrial developments under our present market system, more and more waste material and energy are returned after use to such common property resources as the air, the water, complex ecological systems, large landscapes, and the electromagnetic spectrum. These resources, for which there is no market-exchange mechanism, are progressively degraded because their use as "dumps" appears costless to the industries, municipalities, and individuals using them. This is so even though their values for other uses are degraded or destroyed. Thus, destruction of fisheries or increased mortalities from respiratory disease simply do not enter into the profit-and-loss calculations of economic enterprises, which get their signals and incentives from market prices.

This phenomenon Garrett Hardin has called the "tragedy of the commons." Because the commons are free, no individual user has the incentive to husband them or protect them. Naturally enough, the commons become increasingly degraded and yield a narrower range of services and values.

III. *The Nature of Antipollution Policy*

Before looking at present and prospective policies for antipollution, I propose to state some principles. The first principle has to do with the choices available to polluters. Actually there is a variety of alternatives for reducing the flow of wastes to the environment. Typically, we think of treating streams to reduce the quantity of residual waste ultimately discharged. But it is also possible, especially for industries, to recycle waste streams for recovery of waste materials for reuse or even for sale. Studies of the sugar beet industry, for example, have shown that immense possibilities exist for recycling and waste recovery.[4] The recent creation of a National Center for Solid Waste Disposal is encouraging; a consortium of major corporations has set it up to collect and disseminate information and to design and carry out projects to solve the solid waste and litter problems.

Technology can effect change at the consumer's end by producing less-polluting products, or at the

4. George Lof and Allen V. Kneese, *The Economics of Water Utilization in the Beet Sugar Industry*, distributed by the Johns Hopkins Press for Resources for the Future, Inc. (Washington, D. C., 1968), 125 pages.

producer's end by employing less-polluting production processes. Internal-combustion engines with lower emissions of harmful substances, or consumer goods with lower disposable fractions, are examples of products that can be made at the consumer's end, while production processes which make more efficient use of fuels and other raw materials to produce smaller quantities of wastes are examples of changes that can be made at the producer's end. Or wastes can simply be contained, as is the practice with radionuclides (though this poses problems of safekeeping).

Social policy can help steer consumption and production away from polluting goods to less-polluting goods. This may mean either slowing the rate of increase in consumption of, for example, automobiles and electricity, or reducing the fraction of polluting goods which make up the total consumption budget, or both. This is the ultimate thrust of the environmentalists' economic concern. We believe that policy has too long fostered growth *qua* growth and production *qua* production with little or no regard for survival factors. Perhaps, the chief sources of social welfare are to be found in a far more selective form of development, as Mishan suggests.[5] Environmentalists would redirect much of consumption away from consumer goods—automobiles, refrigerators, highways—toward the service goods—educa-

5. Ezra J. Mishan, *The Costs of Economic Growth* (New York: Praeger, 1967).

tion, health, parks. This, in Professor Galbraith's terms, would restore social balance,[6] and it would lead in a true sense to a higher quality (standard) of living.

The second principle concerns the efficiency of our efforts at reducing environmental pollution. Our antipollution policies had better be successful in finding that combination of the many alternative ways available for cleaning up the environment that will be least expensive. What the economist and systems analyst call cost minimization should be our strategy if we are to achieve ambitious goals. Were we facing a small problem we could ignore the dictates of efficiency at small peril, but repeated estimates of the costs of reducing environmental pollution make it plain that we are facing expenditures of between 50 and 100 billion dollars in the next five years, or between 1 and 2 percent of the total GNP during that five-year period, if we are to achieve substantial reduction in environmental pollution.[7] If, we are inefficient in the way we select anti-pollution methods and put ourselves in line to pay twice what it should cost to achieve these goals, then our losses will be large either in terms of additional resources required or in terms of failure to make significant improvements in environmental quality.

6. John Kenneth Galbraith, *The Affluent Society* (Boston: Houghton Mifflin Company, 1957).

7. Kneese, *op. cit.*

The third principle concerns the standards of environmental cleanup which we wish to aim for. We all assume we want a clean environment, but how clean is clean? What portion of our consumption goods are we willing to forego in order to cut down on environmental pollution? We may again borrow a term from economics and systems analysis and talk about the "optimum degree" of pollution, but while this can be defined conceptually, there are few who believe that we are close to defining it in any practical situation. What we can do, however, is to adopt what Professor Wantrup has called the "safe minimum standard."[8] With air pollution, this might mean air which did not endanger human health. With toxic substances this may mean setting tolerance levels at or near zero. In aquatic environments, this would mean levels which did not irreparably damage the ecosystem. We would not expect the same minimum standard to equal the social optimum, but if we were observant, we could detect the costs and consequences of achieving such a level and then decide on further directions and magnitudes of change. Our guiding principle in these decisions can be, as the President's Council of Economic Advisors suggests, that we want to eliminate pollution when the physical and aesthetic discomfort it creates, and its damage to people and things, are more costly than

8. S. V. Ciriacy-Wantrup, *Resource Conservation: Economics and Policies*, 1st ed. (Berkeley: University of California Press, 1952).

the value of the goods whose production or use has caused the pollution.[9] Not all production pollutes, and not all production is essential to public well-being, so we can afford to be choosey.

This incremental attack on the problem of deciding how much pollution is tolerable will at least lead us in the right direction and—this is important—if pursued resolutely, can lead us rather quickly, and much less painfully than some think, to vastly improved levels of quality in our environment.

IV. *The Choice of Instruments for Antipollution Policy*

If we are agreed on the wisdom of finding the most efficient combination of antipollution measures, that is to say, the combination which achieves at least a safe minimum standard at the least social cost, then considerable study and ingenuity are justified in selecting the instruments for that policy which will put that combination of measures together.

An "instrument" in these terms is a device which causes a producer or consumer to change his behavior in a way which lessens environmental pollution. Russell Train has pointed out that the range

9. *Economic Report of the President* (Washington: U. S. Government Printing Office, 1971), Chapter 4.

of policies and institutions to control environmental abuses is very broad.[10] They include:

> government standards for products, such as standards for automobile emissions
>
> government standards for production processes, such as limits on the discharge of effluents into rivers by factories
>
> taxes on emissions or other substances causing pollution
>
> subsidies for control of pollution through tax rebates, or payments to industry to offset costs of pollution control, or government expenditures on projects for the improvement of the environment

A review of existing pollution programs reveals that our strategies have been incomplete. We have relied heavily on government standards for production processes by attempting to enforce, through regulation, limits on the discharge of pollutants into the environment. This program has been supplemented by subsidies in the form of tax rebates to industry and grants to municipalities and other public bodies to assist in meeting the costs of pollution control. More recently, we have begun to use gov-

10. Remarks of Russell E. Train, Chairman, Council on Environmental Quality, before the Atlantic Council —Batelle Memorial Institute Conference, Washington, D. C., January 15, 1971.

ernment standards for products, as in automobile-emission standards.

By any measure, the progress of our programs has left much to be desired. The General Accounting Office in a recent report to Congress has revealed that there is only disappointing progress in the program to give grants for the construction of municipal sewage-treatment works and notes that even in the works which have been constructed, high operating standards are not enforced.[11] The national press is full of reports of delays in fighting pollution. Recently it was reported that in the Lake Erie basin, 78 cities out of 110 had failed, most by more than a year, to meet original deadlines set at a pollution abatement conference. Industry, too, was reported lagging; 38 plants were more than a year behind in their abatement programs.[12]

We are beginning to realize, I believe, that the instrument for enforcing regulation, with or without the aid of subsidy, offers a cumbersome and inherently slow means of achieving antipollution goals. There are many reasons for this, but the central one is that each enforcement action is an individual action requiring its own set of administrative and

11. General Acounting Office, *Need for Improved Operation and Maintenance of Municipal Waste Treatment Plants.* Report to the Congress by the Comptroller General of the United States, Washington, September 1, 1970.

12. *Wall Street Journal* (March 4, 1971).

legal resources and offering polluters the opportunity of too many kinds of delaying tactics.

Subsidies, such as the current grants to municipalities for waste-treatment plants, have a major defect in that, typically, they are tied to specific measures. This, in turn, usually leads to a violation of the least-cost principle of waste reduction because polluters are not encouraged to make free and unbiased choices among the alternatives.

The grants program for the construction of municipal waste-treatment plants has suffered particularly from the pernicious effect of industrial wastes. They are being discharged at increasing rates into municipal sewers at free or nominal rates, which is disrupting the operation of the municipal plants. The GAO investigators cited particularly disruptive waste discharges from poultry plants, abattoirs, and tanneries. In this connection, I am pleased to see that the President's program calls for a charge to be levied by municipalities against industrial waste discharges in order to recover the costs of treating those wastes. This should encourage pretreatment, recycling or withholding of waste by industries which heretofore have been taking advantage of the municipal grants program in dumping their wastes.

It is even more encouraging that the President and his advisors on environmental affairs have shown great interest in adding to our overall arsenal of antipollution instruments by employing direct charges against polluters as an incentive to get waste loads reduced. In favoring charging industry for wastes discharged rather than making subsidy payments, the administration has concluded, correctly, in my view, that the principles of equity and fairness are better served if polluters are charged for pollution rather than paid to reduce it. The President, in his 1971 State of the Union message, enunciated the principal argument for this position:

> We no longer can afford to consider air and water common property, free to be abused by anyone without regard to the consequences. Instead, we should begin now to treat them as scarce resources, which we are no more free to contaminate than we are free to throw garbage in our neighbor's yard. This requires comprehensive new regulations. It also requires that, to the extent possible, the price of goods should be made to include the costs of producing and disposing of them without damage to the environment.

The arguments in favor of charging industries that pollute may be summed up in terms of incentives and revenue raising. The latter point needs little comment. Greater reliance on charges rather than expenditures from the central treasury not only lightens the burden on the treasury but also provides a source of revenues to be used for collective antipollution facilities such as regional waste-treatment plants, or for recreational facilities or other means of mitigating the damages of pollution. Estimates in connection with Senator Proxmire's bill to create a national effluent charge on water polluters suggested that revenues would come to 2 or 3 billion dollars annually.[13] The point to be emphasized is that rather than draining the central treasury, antipollution charges will supplement it. But, of course, they represent a source of revenue we should be glad to see dry up and which would dry up if rates were set to create incentive.

The incentives to be gained from a system of antipollution charges have been elaborated by a number of economists, but most notably by Kneese.[14] The workings of a charge can be seen in the example of the President's proposal for a sulfur emissions tax. This emissions charge would be based on a calculation of damages caused by the dispersal of sulfur oxides in the atmosphere. These include national health costs, property damage, and damage to vegetation. These costs are estimated at 8.3 billion dollars or about 20 cents for each pound of sulfur emitted

13. Sanford Rose, "The Economics of Environment," *Fortune* (February 1970), p. 186.

14. Allen V. Kneese and Blair T. Bower, *Regional Water Quality Management* (Baltimore: Johns Hopkins Press, 1968).

into our atmosphere.[15] Although the charge has not yet been set, we would favor 20 cents per pound, if this is the value of the damages. The charge would be levied on the sulfur content of the fuels burned by power plants, smelters, refineries, and others, with rebates given according to the proportion of sulfur removed from fuels or recovered from the stack gases.

The incentive effect of the charge is straightforward. The polluter can escape the charge in part or *in toto* by reducing or stopping his sulfur emission. If he found he could reduce sulfer emissions for less than 20 cents per pound, he would do so. If one plant found that a complete reduction paid, while another could only reduce emissions by 40 percent before costs of the next step in reduction exceeded the charges any further reduction avoided, this outcome, while far from uniform, would nonetheless lead to an effective reduction of sulfur pollution.

The emission charge voids the familiar argument which so often is used to delay the imposition of enforcement actions against polluters, that is, that the technology for achieving a given control is not available. By decreeing that use of the environment for waste disposal is no longer free, we require a payment for pollution. If the polluter can possibly

15. *The President's 1971 Environmental Program: Controlling Pollution,* Book One of a Three Part Series, Executive Office of the President, Washington, 1971.

find or develop a technology to control wastes, the charge gives him every incentive to do so. Although regulatory cases often drag on for years, no one argues with the tax collector.[16] In the case of sulfur dioxide pollution, we appear to have little time for argument. The President's advisors tell us that, by the year 2000, sulfer dioxide emissions are likely to quadruple unless action is taken to reduce them.

We may not always come up with an estimate of the damages of pollution as handily as we came by the sulfur damage estimates, but this does not preclude the use of antipollution charges. By resorting to a safe minimum standard and then calculating for an average or typical polluter the costs of meeting that standard, we could set a charge which would give polluters incentive to meet the standard. Suppose the standard called for a 90 percent reduction in emissions and a study showed that this would cost 2 cents per pound of finished product to achieve. If a charge were applied of 3 cents per pound of product, then with a 90 percent control, a producer would save 2.7 cents of the charge per pound. If indeed it costs 2 cents per pound to achieve that 90 percent control, the producer would

16. The impersonal nature of a tax deserves more emphasis. It is reported by Robert H. Boyle that a member of one of the enforcement agencies has excused slow action with the comment "We're dealing with top officials in industry, and you just don't go around treating these people like that." *Audubon* (March 1970), p. 48.

not hesitate to do so, since he would thereby avoid the larger charge. If, as likely would be argued if enforcement were used, the 90 percent control were "technologically impossible," the producer would engage in less-complete control, but he would have strong incentive to search for economical improvements in his control technology.

The antipollution charge would yield information which would serve to evaluate the safe minimum standard. If it appeared it could be achieved at a very low cost, we might consider raising the standard to a higher level of environmental quality. If it appeared very costly to reach, far more costly than the charge imposed, we would know that the particular environmental cleanup we had in mind was going to incur far more serious costs than we had envisaged, and we would reassess our choices. In either case, we would have a greatly improved information base for assessing whether a particular price was high enough so that the costs of the antipollution measures were justified by the improvement in quality of the environment. In this fashion, the information generated by an incentive charge would help us to determine whether the safe minimum standard chosen was anywhere near the social optimum.

V. *The Burden of Antipollution Programs*

It is abundantly clear that environmental pollution is costly. While I will not attempt an exhaustive analysis of these costs, I have cited sufficient evidence to make the point.[17] We have also pointed out that these costs are seldom borne by the polluter, although he too may suffer from pollution created by someone else. It is equally clear that if the environment is to be cleaned up, the cleanup, too, will entail substantial costs. The estimate of 50 to 100 billion dollars cited earlier refers to the total private and public budget for the next five years and has been culled from a number of sources in a useful analysis by Jane Brashares.[18] For a number of reasons the magnitude of costs to be borne initially by industries for antipollution control is particularly ambiguous. To our knowledge no complete estimate of the industrial costs of antipollution programs has been made, or can be. The most extensive estimate of industry expenditures has been compiled by the National Industrial Conference Board. Their figures suggest that the manufacturing industries spent 1 billion dollars in capital for pollution con-

17. For an excellent summary of these costs see *Fortune* (February 1970).

18. See Appendix B in Kneese, *op. cit.*

trol in 1968. The electric utilities are expected to spend from a half to 1.4 billion dollars per year to control sulfur dioxide, or the equivalent of 10 to 20 percent of the cost of generating power. General industrial abatement of water pollution could cost 3 billion dollars over the next five years for equipment plus a half billion dollars or more for operation.

While these examples serve to suggest ranges of costs we may confront, ambiguities stem from the fact that many and often the most efficient antipollution remedies are themselves productive expenditures in terms of by-product recovery or input and process up-grading. Therefore to charge the full cost to pollution control is misleading, and any cost apportionment would be highly arbitrary. We suspect that if antipollution measures receive the deserving attention of industrial ingenuity, they will prove to be much more beneficial to industrial production and less of a cost item than now appears likely.

Although we are not insensitive to the matter of the total costs of antipollution, we entertain no doubts that quite large costs, probably of the magnitudes being discussed, are justified in the name of antipollution. We are concerned not only with the size of the bill but also with the problem of who should bear the costs. While it may be a truism to say that the public ultimately will bear the costs either in increased taxes or in increased prices for consumption goods and possibly in unemployment and reduced output, this misses many of the subtleties of the question of burden. In many cases, the initial impacts are as important as the ultimate impacts.

In distributing the burden of antipollution costs, there are essentially two groups in society who can pay. One is the polluters, and the other is the public at large—those who are polluted. At times, we are polluters, as when we drive or ride in motor vehicles. Whether our government subsidizes the production of nonpolluting motor vehicles, or requires the production of such vehicles with the manufacturers forced to cover costs (almost certainly to be passed along to consumers) might not make radical differences in the distribution of the burden. Similarly, if a sulfur-emission tax raises the price of electric power, most of us will pay for this change, whether it is financed as a subsidy or as a consumer charge.[19] But in other cases where the pollution is from a source such as a large factory

19. We are skirting obvious issues here which seem to lead into evermore complex issues. For example, subsidizing sulfur control in power generation through Federal taxes will produce a different distribution of the burden among power users than if costs are passed along as higher prices. This is because Federal taxes are progressive with income while charges for electricity are probably regressive. Moreover, costs of sulfur control will vary among regions so that some market areas will be less affected than others while the Federal income tax might have a different regional distribution.

of a particular city, and a specific group of people are the customers or residents, as the case may be, then there will be sharp differences in the distribution of the burden depending upon whether antipollution is subsidized by the government or is paid for by the polluters.

Let us accept the prevailing view that equity and fairness are served by imposing the costs of antipollution on the polluter. The nature of these burdens, that is, whether polluters are to be charged for emissions and encouraged to find their own least costly solution or are required to pay the costs of reducing emissions to conform to emission standards, will be such as to affect the costs of production. What happens beyond that depends on corporate behavior.

Discussions of the shifting and incidence of business-tax burdens are a respectable part of public finance that have been carried on for a long time. One, therefore, is inclined to view the imposition of pollution control costs on a firm in the nature of a tax. The question is what sort of tax is the analog of pollution control costs? An emission charge or a required reduction in waste emission might be considered the equivalent of a tax on unit of output on the grounds that waste production is a linear function of output. If required investments in waste reduction technology are viewed as a burden on profits, then pollution control costs may be viewed as a profits tax. The theories and empirical evidence of reactions to unit taxes are perhaps a bit nearer unanimity than the theories and evidence of behavior to the corporate tax.[20]

The general view of a unit tax would seem to be that it is incorporated in the price of the product on which it is levied and has an effect on both price and output. If corporations tend to view antipollution costs as a tax on profits, then behavioral theories are much less clear and the evidence that the burden is shifted to the consumer is conflicting.[21] The best that one can say would seem to be that over the long run, and during a period of general equilibrium, the costs of antipollution will affect prices, profits, and investment in production in some combination of ways. We will briefly discuss the simple case of an increase in prices, then some evidence on the magnitude of cost effects, and, finally, the more complex case of general equilibrium adjustments.

It is worth noting that if polluters pay the costs

20. Richard Musgrave, *Public Finance* (New York: McGraw-Hill Book Company, Inc., 1959), reviews the state of economic understanding of the effects of business taxes. M. Krzyzaniak and Musgrave, *The Shifting of the Corporate Income Tax* (Baltimore, 1963), and R. J. Gordon, "The Incidence of the Corporation Income Tax in U.S. Manufacturing, 1925-1962," *American Economic Review*, Vol. 57 (September 1967), pp. 733-758, are more recent contributions.

21. See Gordon, *op cit.*, and communications by Krzyzaniak and Musgrave and Gordon in *American Economic Review*, Vol. 58 (December 1968), pp. 1358-1367.

of meeting standards or are charged for pollution, we can expect the prices of their products to increase. This increase will serve as a signal to consumers to adjust their consumption away from products that cause pollution. If, however, polluters are subsidized to reduce pollution, then total private consumption may be reduced because increased taxes would have reduced incomes. But consumption can be steered away from the products that cause pollution only by imposing further regulations or sumptuary taxes. The lesson would seem to be that if we are to restore consumer sovereignty as a guiding principle, then wisdom dictates that the costs of goods be made to reflect pollution costs. The polluter should pay for the sake of social control as well as of equity.

If the burden of antipollution programs is borne by polluters, fears sometimes arise as to the impacts of these added costs on various industries, companies, and plants. Will the burden close down plants? Will companies be ruined?

Will the competitive position of American industries in international trade be upset? Fortunately, some evidence exists on these points, and it does not appear to be terribly bad news at all.

Perhaps the most detailed study of the costs of pollution control is the Delaware Estuary Study which was completed by the Federal Water Pollution Control Administration in 1966. This study examined not only the costs of achieving different standards for water quality but also the costs of different administrative programs for achieving a given standard.[22] It concluded that considerable savings could be realized if the mathematically programmed solution to achieve the least cost could be approached in real life through appropriate programs. The usual approach of assigning uniform levels of treatment to all polluters produced dismal results, in this regard; it cost roughly three times what it would have cost if manufacturers were allowed to seek a solution that would have resulted in a minimal cost if the favored objective of achieving 3 to 4 ppm of dissolved oxygen were attained.

It was decided to experiment on paper with incentive charges to see whether rational action to avoid a charge were taken would cause polluters to reduce their discharges so as to achieve the standard at a lower cost than the uniform treatment program would. It was found that a charge of 8 to 10 cents per pound of BOD (biochemical oxygen demand) discharged achieved the standards at roughly half the cost that the uniform treatment program did. These results are important in and of themselves, but this lengthy prelude is necessary in order to understand the point that, as a result of the charges and costs of waste treatment, the researcher was able to estimate the costs of treatment, plus

22. For an excellent recapitulation of this study see Kneese and Bower, *op. cit.*

effluent charge for each industry.[23] Table 1 reproduces the results. It is noteworthy that in all but a few cases it was found that the total cost was less than 1 percent of the value of output. In most cases, it was a small fraction of 1 percent.

TABLE 1

Industrial Antipollution Costs as a Percentage of Value of Output in Delaware Estuary

Industry Class	Firm No.	Cost of Treatment Induced By 10 Cent Effluent Fee [a]	Total of Treatment Plus Fee
Paperboard mills	1	1.0	2.1
	2	0.0	3.0
	Average	0.6	3.4
Inorganic pigments	1	0.0	0.9
Petroleum refining	1	0.06	0.12
	2	0.08	0.14
	3	0.16	0.22
	4	0.14	0.22
	5	0.03	0.05

23. Edwin C. Johnson, "A Study in the Economics of Water Quality Management," *Water Resources Research,* Vol. 3, No. 2 (1967).

	6	0.09	0.09
	7	0.00	0.38
	Average	0.08	0.13
Industrial			
Organic chemicals	1	0.18	0.43
	2	0.00	0.33
	3	0.31	0.31
	Average	0.25	0.32
Organic chemicals	1	4.0	4.0
	2	0.0	4.0
Paper mills	1	0.0	0.34
Building paper board mills	1	0.20	1.2
	Average	0.13	0.26

a. Effluent fee of 10 cents per pound of BOD (biochemical oxygen demand) discharged.

Further evidence on the costs of waste control by industrial groups has been assembled by Kneese. His estimates are based on stringent control measures and our present knowledge of waste-control costs.[24] The estimates shown in Table 2 reflect best guesses. Kneese believes that the lower values shown represent most likely values while the upper limits represent the worst that could happen to an industry or to a few firms in that industry.

24. Kneese, "The Economics of Environmental Pollution in the United States," *op. cit.*

TABLE 2

Estimated Costs of Stringent Pollution Control as a Percentage of Current Costs of Production

Industry Group	Percentage
Foods and live animals	1–5
Beverages and tobacco	1–5
Crude materials	1–10
cotton and soybeans, hemp, etc.	1–5
metal ores, paper base stocks, textile fibers, rubber, concentrates	5–10
Mineral fuels, petroleum products	5–10
Oils, fats, waxes	5–10
Manufactured goods	5–10
textiles, iron and steel, nonferrous metals	5–10
Machinery and transport equipment	5–10

Source: Edwin L. Johnson, "A Study in the Economics of Water Quality Management," *Water Resources Research,* Vol. 3, No. 2 (Second Quarter, 1967), Table 8. *Source:* Allen V. Kneese, "The Economics of Environmental Pollution in the United States," *op. cit.*

The most notable thing about Kneese's estimates is that the extreme of the increase in costs of production due to the imposition of strict waste control programs is no more than 10 percent in any industry, and the most likely value for a majority of the industrial groups is 5 percent. The agriculturally based industries, food, beverages, tobacco, soybeans, and fibres, get off lightly with an estimated 1 percent increase in costs. Considering that perhaps five years would be required to effect fully the "stringent" programs on which these estimates are based, a 10 percent increase in five years is equivalent to a compound rate of slightly less than 2 percent per annum while the 5 percent increase would be equivalent to slightly under 1 percent per annum. These very modest rates of change are the sort of magnitude of relative changes to which our economy adjusts constantly.[25]

Ultimately these increases in costs, the changed nature of investment, and the effects on profits will affect the equilibrium position of the economy. Although this problem is too intricate to be reasoned out, it is possible to model the economy mathematically and to stimulate the model for the purpose of estimating the consequences of altered investment and price behavior which might occur with stringent pollution control. Dr. Robert Anderson[26] has performed this analysis on air-pollution control by assuming anti-air-pollution investments of 1.2 bil-

25. As an example of the magnitude of the relative change in the economy, output per man hour has grown by 8 percent since 1967; unit costs of forestry output have grown at 0.9 percent per year for an 87-year period; from 1919 to 1956 unit cost of minerals declined 3.2 percent per year; since 1939, the proportion of the population in farming has decreased at the rate of 4 percent per year.

26. Reported in Sanford Rose, *op. cit.*

lion dollars annually by industry, 320 million dollars annually by the electric utilities and an increase of 1 percent in new car prices stemming from exhaust emission controls.

His results, applied to the economy between 1962 and 1964, dropped GNP from an annual rate of 625 billion without pollution controls to 617 billion dollars with air-pollution controls. The loss of 1.3 percent in GNP we might feel to be more than offset by relief from the real costs of air pollution so that a net welfare gain could be quite consistent with a reduction in G.N.P.[27]

Anderson's conclusions about the effects of unemployment are not so easily dismissed. He found that unemployment would rise from 4.8 percent to 5.3 percent. This half-a-point rise in a work force of 80 million means 400,000 additional jobless. To them the burden of antipollution would seem onerous indeed and to the extent that existing compensation, retraining and relocation programs were inadequate or inequitable, these unemployed would deserve special attention. But while one wishes to avoid callousness toward the unemployed, an effect of this magnitude, particularly if it were temporary, does not seem to be an unreasonable price to pay for air-pollution control. We may also hold out hope that the freedom we would give industry to find its

27. See Mancur Olson, "The Plan and Purpose of a Social Report," *Public Interest* (Spring 1969), pp. 85-97, for more development of this point.

own least-cost solutions to its pollution problems would work to make these national economic effects even smaller than Anderson's conclusions suggest.

Many industries produce products used by other industries. The chemical industry is a good example. A rise in the costs of products sold by one industry to another will not necessarily mean a rise in the costs of the production processes in which they are used. If the industries that purchase products from other industries can find substitutes at no increase in costs or can adjust technology to offset the increased costs of some inputs (by using less), costs need not rise. It must not be overlooked that as industries move into antipollution programs in a substantial way, a new set of demands between industries for waste-reduction techniques and equipment will be generated. This will create, among other things, new business opportunities. Putting all of these effects together and coming to conclusions about the ultimate effects of pollution control the relationships between industries is a task for which there are analytical techniques.[28] We can only urge the Federal government to take an interest in such calculations.

The anticipated effects of pollution control on the domestic economy do not necessarily lead di-

28. Wassily Leontief, "Environmental Repercussions and the Economic Structure: An Inout-Output Approach," *Review of Economics and Statistics,* Vol. LII, No. 3 (August 1970), pp. 262-271.

rectly to conclusions regarding the effects of antipollution programs on the position of the United States in international trade. This complex problem was investigated by Professor d'Arge,[29] who finds little to fear from adverse affects on international trade from the imposition of stiff environmental improvement programs. Using the increased costs shown in Table 2, and assuming that the United States will impose environmental controls unilaterally, d'Arge finds that domestic income could drop by 5.2 to 12.6 billion dollars or from 0.6 to 1.4 percent as a result of changed competitive relations. The balance of payments would also be affected, but, surprisingly, it would increase by from 2 to 6 billion dollars because imports as well as exports would be affected. Since the increases in costs forecast by Kneese are not likely to occur instantaneously, d'Arge forced through his model a 1 to 2 percent increase in export prices and found domestic income decreased by only 0.16 to 0.32 percent. This range in the annual differences is more likely to be expected in income while a stiff antipollution program was being put into effect. Although d'Arge's results

29. Ralph D'Arge, "International Trade, Domestic Income, and Environmental Controls" in Kneese, "*The Economics of Environmental Pollution*," *op. cit.* See also his "Essay on Economic Growth and Environmental Quality" to be published in the *Swedish Journal of Economics*.

are rough and ready, they certainly reveal no cause for alarm.

While the foregoing figures on the rise in domestic costs and the loss of domestic income that American industry would suffer from international trade are reassuring, we ought not ignore the fact that different plants and firms in different regions will be affected differently. It would be shortsighted to allow concern for a few extreme cases to delay us from pursuing the changes demanded to put strong antipollution policies into effect, but we should be ready to provide assistance to workers and firms who are particularly disadvantaged by these adjustments in costs and competitive relations. In the case of international trade problems, Chairman Train has suggested that the provisions of our Trade Expansion Act could be extended.[30]

Multinational environmental controls are also a possibility. In fact, d'Arge's figures cited earlier suggest that Japan, West Germany, the United Kingdom, and France, the only other countries studied with his model, would all experience much greater adverse effects if *unilaterally* stringent pollution controls were imposed on their economies than the United States would experience. The maximum estimates of negative change in income exceed 50 percent for all except West Germany. This surely

30. Remarks of the Honorable Russell E. Train, *op. cit.*

provides a strong case for concerted multinational movements to control pollution.[31]

Some plants and even firms will go out of business in the face of strong antipollution policies. However, we need to remember that mortality has always been a fact of life in the competitive business world and that the firms or plants that can be shut down by a 5 or 10 percent increase in costs are probably already not far from being dead. Moreover, considering the sensitivity of the large, multiplant firm to community relations, we can be assured that there are a number of obsolete plants in one-industry towns around the nation just waiting for an excuse to close. Environmental blackmail is an old game, and environmental scapegoating will be the name of a new game.[32]

The advantages of locating a plant in one region rather than another will also shift as antipollution policies take their full bite, and the regions with fewer pollution problems, better disposal technology, or better grades of raw materials will automatically realize these advantages. Our national policy already calls for a stop to degrading water quality any further, so we cannot (and should not) allow pollution merely to equalize costs between regions. Some localities have more serious problems than others,

31. Professor d'Arge is developing the implications of simultaneous multinational pollution controls.

32. See "Pollution Control Layoffs Will Be Probed by Muskie," *The Washington Post* (February 13, 1971).

due to long neglect, high rates of activity, and varying capacities for different locations to assimilate pollution. Trying to neutralize these differences would defeat the desirable effect of allowing these differences to encourage greater rates of settlement and development in the more environmentally advantaged regions. We may wish, in any case, to encourage a decongestion of our congested, environmentally degraded areas.

Tax writeoffs have been a favorite remedy for ailing firms but we ought not to be deluded by their apparent advantages. Writeoffs help only the profitable firm; they do not help the firm which has no profits to tax. Moreover, tax writeoffs or accelerated depreciation allowances encourage installation of waste-treatment equipment rather than recycling and process change, thus moving us away from rather than toward the program that cuts down pollution at the least cost for industry.

VI. *A Note on the Problem of Pesticides* *

Before concluding, I should like to comment briefly on the implications of the cost of environmental protection with regard to that most con-

* This section was largely drafted by Roland C. Clement, Vice President/Biology, National Audubon Society.

troversial of all environmental pollution problems, the use of synthetic chemical pesticides in agriculture, forestry, and elsewhere—an issue, by the way, with which the National Audubon Society has had considerable experience.

It is a tragedy that this controversy still has to go on. It constitutes, in our view, a flagrant case of placing corporate profits above and against the public interest, for industry has been aggressive not only in selling these products, but in waging, sometimes viciously, a stubborn battle against the mounting accumulations of scientific evidence of environmental damage.

Today's farmer is a big businessman, often conducting a million-dollar operation. Like all businessmen he must weigh his inputs carefully, since his profit margins will be affected by these investments. Unlike some other businessmen, however, he has not had to weigh the "spillover effects" of his practices, nor does he appear to care about externalities or neighborhood effects such as environmental poisoning and wildlife decimation. It perhaps would not be fair to expect many individuals to modify pest-control practices unilaterally as long as competing farmers or firms are allowed to use these seemingly cheap control methods and thus undercut them in production costs. Only governmental regulation, via the legal control of pesticides and their use, can eliminate such competitive advantages and place producers in a given area on a reasonably equal footing.

Some of the more emotional apologists for maintaining the status quo in pesticide use have confused both the Congress and the people of the United States by making claims that eliminating pesticides would increase food production costs by 200 percent or more. This overlooks (deliberately?) the point that none of us in the conservation movement has ever asked for the elimination of all chemical pesticides. We merely advocate the elimination of those long-lived, fat-soluble materials that are cycled, and cumulatively concentrated, in nature's food webs—and of those containing long-lived heavy metals, like mercury, whose toxicity is just too great to put up with. In any event, the only solid economic evidence we now have along these lines is that derived from the British experiment in greatly restricting the use of aldrin, chlordane, dieldrin, DDT, endrin and heptachlor.[33] This, it was forecast, would probably increase agricultural production costs by about 3.5 percent, and only for that period of time necessary to perfect alternative control measures. The British restrictions were put into effect in 1964; in 1966 they sprayed only 327,300 acres of their total 10.4

33. Reported on by Roland C. Clement, "The Pesticide Problem," *Natural Resources Journal,* University of New Mexico School of Law, Vol. 8, No. 1 (January, 1968).

million cropland acres with these chemicals,[34] and we have certainly not heard of a collapse of that nation's agriculture.

The potential impact, on either corporate profits or the food supply of the nation, of changes in pest-control measures calls for a sophisticated analysis of the alternatives open rather than the taking of extreme all-or-nothing positions. There are several pest-control strategies, some of which are not dependent on chemicals at all. Where chemicals seem honestly to be called for, there are usually several alternative and effective chemicals available, some of them quite acceptable to conservationists. In fact, the National Audubon Society annually publishes a list of the more readily available pesticides acceptable to conservationists in the control of common insect pests.

Let us remember, also, that seldom is one crop threatened throughout the area in which it is planted by any single pest. This means that even if no controls were used, the damage done would be local or regional and could be made up by increased production elsewhere. We are, in short, really involved with questions of production at the margin —not an all-or-nothing situation—and with ques-

34. Report by the Advisory Committee on Pesticides and Other Toxic Chemicals, "Review of the Present Safety Arrangements for the Use of Toxic Chemicals in Agriculture and Food Storage," London, Her Majesty's Stationery Office, 1967.

tions of local disaster which have always been with farmers. The millions now invested in pest control have pushed the farmers' costs of production sky-high. For example, many producers will invest 12 dollars in pesticides to produce a single bale of cotton. By limiting pesticide use, we would favor production in those regions where the pests are much less troublesome, and both the real costs of production and the real costs of raw materials for the nation's economy would actually decline. This is very close to a situation where almost no one except the chemical industry and the adversely affected farm regions are worse off and most of us are better off, and it is to this that the conservation movement has been directing its efforts on this problem.

VII. *Conclusions*

In this paper we have attempted to deal in a small space with a large problem. We have had to omit the treatment in any depth of a number of important issues. Among these is the role of the new permit system, under the 1899 Refuse Act, on which the Federal government intends to press for the cleanup of inland waters. Also, the proposal to issue users certificates for ocean dumping raises a whole new range of possibilities for using property incentives to reduce and control pollution. The implications of antipollution policies for international trade are deserving of a good deal more attention, as are the implications of an effective antipollution policy for the composition of our domestic goods and services. Issues of national economic growth and national population growth were largely outside our assignment, but they are certainly germane to full discussion.

In reaching the central conclusion that economic incentives ought to be added to our arsenal of antipollution weapons, we are not proposing that we tell the regulators to stop making and enforcing regulations. We are suggesting that there are important tools which have heretofore been neglected and, equally important, that there are considerations concerning efficiency which ought to permit polluters a wide range of choice in minimizing pollution and its costs.

The nature of our antipollution programs can be what we choose to make it. We have emphasized that there is considerable latitude in the choice of instruments we can make to construct an antipollution program; but rather than describe specific program details, we have held out for the principle that we ought to seek that combination of instruments and policies which will maximize the effectiveness of antipollution expenditures. We have left no doubt, we trust, that large expenditures are justified for antipollution. It is because the stakes are so high that we have chosen to emphasize the principle of efficiency or effectiveness in our antipollution efforts.

Government will, of course, share a certain amount of the costs of antipollution programs in the form of administration, monitoring, and research. But in this paper we have been more concerned with the costs of antipollution programs to the private sector and, in particular, to the corporation. Although the costs faced by many corporations will increase as pollution loads are cut back, and as polluters pay damage charges for residual pollution loads, no evidence has been found that alarming increases in production costs will be experienced by any industry. In most cases, the cost increases appear to be on the order of 5 percent or less. There will be individual cases of hardship which will merit relief, but we ought not compromise an entire program in order to save a few hardship cases. We need more direct evidence on the issue of corporate profits, but the best way to get this may be through experience with economic incentives for antipollution.

A really effective antipollution program, in which economic incentives play a proper role, can be expected to change the mix of goods and services produced in the domestic economy, but we applaud it as a move away from products and processes that are heavily polluting to others that are less so. The economy will move to a new equilibrium of production and employment. In this area and in the area of the effect on international trade, we deserve more definitive predictions of the consequences of moving into an age of antipollution; but the evidence we have is that there is no reason for any paralyzing fear that revolutionary consequences will confront us.

The best policy, we believe, is to move ahead with a vigorous antipollution program, but with eyes and minds open to the consequences both in costs and effects, so that, by being flexible, we can improve as we go along both the goals we set for environmental improvement and the effectiveness with which we attain those goals. This, we submit, will require some modifications in posture, on the part of both the antipollution agencies and the polluters.

ECONOMIC GROWTH AND THE PROBLEM OF ENVIRONMENTAL POLLUTION

Solomon Fabricant

WE ARE NOT going to settle, here and now, the very important issues raised in this discussion over the problem of environmental pollution and its relation to economic growth. Our objective, I presume—as is appropriate in a university setting—is to exchange ideas and views. These we will take home to digest and to use in considering, or reconsidering, our positions on what should be done.

Let me give my own ideas on what should be considered if we are to work our way through to sensible decisions. If what I say duplicates what has already been said or implied, view it as putting emphasis on what is important. If it differs, perhaps it will serve to redress any imbalance there may be in the discussion so far.

The problem of environmental pollution is, of course, raised by life itself. Put in the general terms that biologists like to use—I quote Professor Philip Handler and his collaborators—"living things are systems that reproduce." This is also to say that living systems maintain themselves, at least long enough to reproduce; and that they maintain themselves, and the species of which they are members, by using materials absorbed from their environment. These materials are returned, in modified form, to the environment as waste products of the things that continue to live, and as the remains of the things that die.

More specifically, human life means garbage and sewage and exhaled air. And the greater the number of human beings and the higher the standard of living attained by them, the greater tends to be the volume of undesired by-products of production and consumption—"residuals," to use Dr. A. V. Kneese's euphemism—that is returned to the environment. This is why environmental pollution is a more serious problem for us today than it was for earlier generations, and why it threatens to be even more serious for the generations ahead.

To be still more specific, consider some figures: During the lifetime of many of us here, say over the past half century, the population of the United States has about doubled, and real national product per capita has been multiplied by two and a half. The physical volume of production and consumption, then, is now five times what it was around 1920. But the space in which we live is no wider. Further, looking fifty years into the future, to the year 2020, we may expect continued rapid growth from our present high level. Even if nine out of ten women now starting their child-bearing aimed merely at having only two children, and the other one of the ten, only three—which would average to births at the replacement rate—and net immigration were to be stopped entirely, population would still rise by as much as 65 million, or a third. But it could also rise much more: Population could again double, as one of the Bureau of the Census projections indicates. And if the national product and consumption per capita at full employment were to rise at the rate that has been projected for 1970 to 1975, which is 3 to 3.5 percent per annum, the aggregate volume of production and consumption in the year 2020 would be six, eight, or even ten times what it now is.

I hesitate to scare you with projections for the entire human race, which is now increasing yearly by something like 70 million. The rate of world population growth is well in excess of the rate in the United States; and the rate of increase in goods and services per capita hoped for greatly exceeds the rate the United States has experienced in the past. If we were to assume merely a doubling of the Earth's population—which is below the present rate —and merely the attainment in fifty years of the *present* level of GNP per capita of the *poorer* countries in Europe, which would disappoint the hopes and expectations of most developing nations, we would have a volume of solid, liquid, gaseous, and energy effluents from human living on a scale not easy to comprehend. And the Earth would be no bigger than it is now.

But all this only poses the problem. It does not follow that environmental pollution or degradation would rise in proportion to the national product of the United States or of the world as a whole, or even in proportion to the residues of production and consumption. Nor should we assume that there is no feedback, that the direction of causation runs only from economic growth to pollution.

Environmental degradation depends not only on the size, but also on the concentration of the population; on the volume of production and consumption, as well as on the composition, and on the techniques and habits of production and consumption. Environmental degradation depends, further, on the effort made to limit or otherwise deal with the residues of production and consumption. This effort, in turn, depends on its cost, and on the value placed on the benefits derived from it. For a given population and level of living, the lower the cost and the higher the value of preventing garbage, sewage, and foul air from degrading the environment—to continue with these examples—the less pollution there will actually be. But as I have implied, the level of population, and also of production and consumption per capita, is not unaffected by the level of GNP per capita or of pollution. And the other variables I have listed are also interrelated.

If this model is appropriate to the social system with which we are concerned—and I think it is—we may expect that adjustments of one kind or another will be made, will be induced, to meet the problem of environmental pollution. Economic growth and its associated developments tend, for example, to increase the population by cutting the death rate; but they also tend, although with a lag, to reduce the birth rate. Economic growth tends to increase the volume of effluents; but it tends also to increase our economic and technical capacity to deal with them, and strengthen our desire to do so.

This is not to say that the tendencies to meet the problem of pollution always work themselves out in time and to a sufficient degree. Some special effort on our part is needed—and needed not only to reduce the delay, for resources are scarce. No foreseeable rate of increase in productivity will ever eliminate this scarcity. If we are to meet the problem of environmental pollution, and also our other problems, and at the same time raise the average standard of living at a rate to satisfy reasonable expectations, we must become more efficient than we now are in dealing with the pollution problem. We need, in short, to strengthen and to rationalize the tendencies inherent in the social system. As Justice Holmes once said somewhere, for the inevitable to happen takes some sweat. If this sweat is to be properly directed, we need to know much more than we now know about the facts, and about the technical and economic interrelations between the facts.

It is unlikely, for example, that the relationship between GNP and the residue of production and consumption is a one-to-one, or even a linear, relationship. As I have said, the volume of residues depends on the productive process of which the national product is the end result, and on the composition of the national product. Processes and materials and products vary with regard to the volume of residues per unit of goods and services produced and consumed, as any comparison of natural gas and bituminous coal makes clear. Shifts in the structure of production and in the composition of consumer goods—shifts that are *bound* to take place in a growing economy—will affect the relationship between the aggregates. The rise of the service industries is one example of such a shift. And in a primitive economy, to recall another example, night-soil is a necessary fertilizer; it is not sewage. But we need more than just examples.

To deal sensibly with the problem of pollution, it is essential, also, to have a reasonably adequate idea of what it costs or would cost to deal with it. But this we are only beginning to acquire. Nor are we as clear as we should be on the benefits to be derived, in different circumstances, from lessening this or that kind or quantity of residue. In part, this is a job of developing more and better knowledge on the rate of accumulation of mercury and other toxic metals in the food and water supply, and then in the human body, and on the health effects of different quantities or concentrations, for example. In part, also, it involves somehow polling the people on the utility they would impute to such things as green belts and blue skies and clear waters. It is not a simple job.

Related to this is something to which economists have been devoting a good deal of attention in recent years. It concerns the effects, good or bad, on people of "transactions to which they are not parties," as Professor Sidney Alexander has put it. These so-called "externalities" are at the heart of the problem of dealing with pollution. But economists have made relatively little progress, so far, in measuring the effects. Most of us believe them to be sizable, and some economists believe also that they are growing rapidly in relative importance. But we are not sure. Much more work needs to be done to replace surmise with some reasonably hard numbers, if we are to decide what kind of institutional arrangements can best serve to shift the burden of pollution from those who do not cause it to those who do.

Because the problem of pollution is serious now, and threatens to become far more serious in the future, it is easy to rush hastily into slap-dash, ill-conceived, *ad hoc* policies to deal with it. Undoubtedly some things must be done at once, and we know enough to do them sensibly. But this is by no means true of everything. We should take the time to learn and to teach, to research and to experiment, and to try out policies from which we can retreat if they prove to be futile.

Existing and possible institutional arrangements to deal with pollution are too varied to be discussed here. But some general points bearing on policy must be mentioned.

It is important to realize that antipollution expenditures would not add to the cost of living (or force a reduction of real income). The costs of pollution are already being borne; they are merely omitted from our calculations. Expenditures to avoid or lessen pollution, if not carried to the point where the benefit derived from the marginal dollar spent falls short of a dollar's worth, would *reduce* the aggregate social costs of pollution.

Further, with minor exceptions, these costs are in the long run borne by consumers, not by producers. No legislation or regulation can appreciably alter this incidence. In the short-run, however, antipollution costs imposed on producers are not quickly shifted to consumers. Considerations of equity may then support the case for subsidies in those situations in which the pollution problem is urgent; and for gradualness, in those situations in which the problem is not urgent.

But while consumers bear all or virtually all the

costs of pollution in the long-run, they do not necessarily do so in proportion to their responsibility for it. What policy can attempt to do—this should certainly be one of its objectives—is to shift more of the burden to those consumers who prefer goods and services turned out by processes that result in great amounts of pollutants. This, in turn, suggests the desirability of antipollution policies that work by increasing the relative prices of such products. These policies would help distribute the burden more equitably; they would also help to reduce the volume of such products.

A specific suggestion is that of imposing charges for the use of scarce environmental services. User charges would have several advantages. They avoid forcing on producers and consumers the use of particular means of fighting pollution; rather, these user charges encourage producers and consumers to seek out the most economical means appropriate to their particular technical and economic circumstances. The charges also encourage the invention of still cheaper means of production, and help avoid detailed regulations that—like building codes—can easily become obsolescent, and that in any case must be administered by a large bureaucracy. User charges could be imposed across the board; particular industries need not, and should not, be singled out, since every industry makes use of scarce environmental services.

There are all kinds of unresolved questions about these and other possible social arrangements. Experimentation, with effective use (rather than neglect) of the information yielded by the experiments, is desirable before long-term or large-scale commitments are made.

Because I have stressed the need for caution and gradualness, I must add, as one of the first and greatest students of welfare economics, Professor A. C. Pigou, did in a different context, "that gradualness implies action, and is not a polite name for standing still."

The political problem, which I can only mention, is very difficult. In large part this is because national objectives are viewed and valued differently by different people. We are not all of one mind; not everyone agrees on the relative values of what everyone agrees is "good." This means that we must learn to settle our differences—we must be prepared, having urged our case as strongly as we can, for compromise. To you and to me, who value highly the silence of the forest's glade at noon, and the song of the bird at dusk, the prospect of what we shall have to settle for can be anything but pleasant.

Nor can we take pleasure in knowing that, from our point of view, tastes are bound to deteriorate further in the long years ahead. For the values of future generations will be molded by the world into which they are born, and this world will be very different from ours because of the continued process of economic growth. Country mice and city mice do differ; and there are also generation gaps. If the projections I have cited are anywhere near the mark, the gaps may even widen. They will certainly cumulate. Our descendants will set environmental standards that we would view as intolerable.

This brings me to my last point: We are confronted by a problem broad enough, and permanent enough, to draw us into the realm of science-fiction. If pollution is permitted to worsen over the centuries and the eons, we can nevertheless suppose that life will somehow adapt itself. "Living systems are systems that reproduce," yes; but as the biologists define them, they are also systems "that mutate, and that reproduce their mutations." That is why living things "are endowed with a seemingly infinite capacity to adapt themselves to the exigencies of existence"—even in a cesspool.

But we cannot be certain that it is *human* life that would adapt and survive. Nor, in any case, is this the way for us to meet the problem. We want to change men, to make them better men, but men we can still recognize as men. To the exigencies of existence, we should adapt not our bodies but rather our habits and our institutions—habits and institutions concerning population, which I have scarcely mentioned, as well as all the other factors bearing on the problem of environmental pollution.

I am optimistic that we will do so. The discontent evident in this and other discussions of the pollution problem has already set us on the right path. Nor can we permit ourselves to doubt that we will succeed. As William Blake reminds us:

> If the Sun and Moon should Doubt,
> They'd immediately Go Out.

THE ECONOMIC GROWTH: ITS COSTS AND PROFITS

Martin R. Gainsbrugh

WE HAVE NOT yet begun to adapt national accounts and the GNP concepts to reflect the impact of the recent outlays to improve the environment. When such outlays were small in relative as well as absolute terms, they may not have required special conceptual consideration. But now that they have become greater, there is need to consider what economic and accounting postulates are to be used relative to this form of investment, as distinguished from what we hitherto considered to be investment.

Traditionally, all private outlays for plant and equipment are entered into GNP under gross private domestic investment as "the net acquisitions of fixed capital goods" designed to improve productive efficiency and output per man-hour. Such investment was in turn charged off or depreciated *over the long term* in determining the amount of capital consumption to be deducted from GNP in arriving at net national product.

Should we continue to treat the heavy investment in environmental control—in "nonproductive" items—in the same way as all other capital investment? I put "nonproductive" in quotes to emphasize that environmental investment may not directly increase the capacity to produce goods and services, as was true of past outlays for plant and equipment. Or should we treat environmental outlays differently? Should they now be charged off *as a social cost* of doing business during the course of the year in which the investment is made? Should we remove them from the sphere of long-term investment entirely? That is how jigs and dies are now treated. They are entirely charged off in the course of one to three years. Very little illumination has been cast on this particular question. That again brings us back to the problem of defining net national product. Should all of environmental control investment be charged off directly as a *current social cost* in determining *net* national product, or should such outlays continue to be depreciated over the life of the equipment, as are all other plant outlays?

Our work in the field of capital appropriations has, over the years, clearly revealed that profits after tax, current and prospective, are by far the preponderant determinant of new investment rather than cash flow. Once absolute profits begin to move downward, new authorizations for plant and equipment are quick to follow. Expenditures for plant and equipment are late turners but these, too, respond to profit deterioration as they did in the recession we have just experienced.

In evaluating the manner in which the current and prospective costs of environmental improvement are to be met, therefore, this relationship between profits and investment should not be ignored. Profits are not only depressed currently but they have been trending downward over the past five years. In 1966 after-tax profits totaled 48 billion dollars or 6.6 percent of GNP. In 1970 they totaled only 39.4 billion or 4.0 percent of GNP and in the last quarter they slipped to only 39.1 billion or 3.9 percent of the GNP.

This decline in profits and profit margins is now coupled with the excess capacity that prevails throughout manufacturing. Unless otherwise offset, this could limit the response of industry to environmental improvement. Such investment, when made, may well be, in the light of the prevailing profit squeeze, at the expense of outlays that would otherwise have gone into equipment designed to lower costs or for new products or technology. Corporate income taxes still command nearly half of all profits before taxes. Special tax incentives or sheltered treatment for environmental outlays would thus assure more widespread adoption of protective environmental devices. They would also serve to limit the possibility of such outlays poaching upon funds that would otherwise have been spent to accelerate the upturn from the 1970 recession or to reduce the wage-cost-price pressures that make the present price inflation so intractable.

If I may, I would like to say a word or two about the invaluable contribution the GNP measures and the related national accounts have allowed us to make toward achieving a higher rate of national economic growth and better living standards. The GNP was not developed until the close of World War II. Within that limited time, as history goes, it has clearly been established as the best analytical tool thus far developed by economists to help shape public policy, particularly under the Employment Act of 1946, and to assist in both short-run and long-term forecasting. The system of economic intelligence we have developed around the framework of GNP helps us to shape our economic future rather than be shaped by it.

My tour of duty as a business economist began back in the dismal thirties. The primary economic problem then centered around the business cycle. The system of economic intelligence was most primitive, having been built largely around dead statistics based on the past and completely lacking the central core of national accounts. Few, if any, business enterprises could be run without the aid of such accounting controls as an operating statement or a balance sheet. Not until the mid-1930s did we begin to develop such accounts officially and the initial accounts were primarily concerned with welfare rather than the market. They were largely keyed to analyzing how the national income was distributed among the components of production rather than to the relationship of the nation's product to the prevailing and prospective levels of consumption and investment.

The usefulness of our national accounts for market analysis has been amply demonstrated. In much the same way that GNP was independently derived from national income, although conceptionally closely linked, I would hope that a separate "social account" will emerge from the national and international interest that has arisen in measuring the net improvement in the quality of the economy's performance. Various "social indicators" are already being measured and analyzed by the National Planning Association and others. Ultimately, from such efforts may well emerge a set of accounts that represent the heightened social costs that arise in a highly industrialized (or post-industrial) society. As Edward F. Denison has observed, such efforts involve expressing in dollar amounts the value of the deterioration of the physical environment measures never before attempted "and difficult even to imagine." He strongly recommends a program that we at The Conference Board have already begun to start, to collect information on changes in expenditures for environmental protection that will be made *in the future*. Even if such information does not lead or enable us to change the measure of output, it will enable us to interpret better the changes in output and productivity that we observe in the future as well as to know the true costs of the new programs.